EURO Advanced Tutorials on Operational Research

Series editors

M. Grazia Speranza, Brescia, Italy
José Fernando Oliveira, Porto, Portugal

More information about this series at http://www.springer.com/series/13840

Giancarlo Bigi • Marco Castellani •
Massimo Pappalardo • Mauro Passacantando

Nonlinear Programming
Techniques for Equilibria

 Springer

Giancarlo Bigi
Department of Computer Science
University of Pisa
Pisa, Italy

Marco Castellani
Department of Information Engineering,
Computer Science and Mathematics
University of L'Aquila
L'Aquila, Italy

Massimo Pappalardo
Department of Computer Science
University of Pisa
Pisa, Italy

Mauro Passacantando
Department of Computer Science
University of Pisa
Pisa, Italy

ISSN 2364-687X ISSN 2364-6888 (electronic)
EURO Advanced Tutorials on Operational Research
ISBN 978-3-030-00204-6 ISBN 978-3-030-00205-3 (eBook)
https://doi.org/10.1007/978-3-030-00205-3

Library of Congress Control Number: 2018956154

This Springer imprint is published by the registered company Springer Nature Switzerland AG
The registered company address is: Gewerbestrasse 11, 6330 Cham, Switzerland

Preface

In science, the term "equilibrium" has been widely used in physics, chemistry, biology, engineering, and economics, among others, within different frameworks. It generally refers to conditions or states of a system in which all competing influences are balanced. For instance, in physics the *mechanical equilibrium* is the state in which the sum of all the forces and torques on each particle of the system is zero, while a fluid is said to be in *hydrostatic equilibrium* when it is at rest, or when the flow velocity at each point is constant over time. In chemistry, the *dynamic equilibrium* is the state of a reversible reaction where the forward reaction rate is equal to the reverse one. In biology, the *genetic equilibrium* denotes a situation in which a genotype does not evolve any more in a population from generation to generation. In engineering, the *traffic equilibrium* is the expected steady distribution of traffic over public roads or over computer and telecommunication networks. Even more, the well-known *equilibrium theory* is a fundamental branch of economics studying the dynamics of supply, demand, and prices in an economy within either one (*partial equilibrium*) or several (*general equilibrium*) markets: the basic model of supply and demand is an example of the former, while the Arrow–Debreu and Radner models are examples of the latter.

Actually, the term "equilibrium" has also always been very relevant in mathematics, particularly in dynamical systems, partial differential equations, and calculus of variations. After the breakthrough of game theory and the concept of Nash equilibrium, the term has been used in mathematics in much larger contexts involving relevant aspects of operations research and mathematical programming. Indeed, many "equilibrium problems", including some of them mentioned above, can be modeled in this framework through different mathematical models such as optimization, complementarity, variational inequalities, multiobjective optimization, noncooperative games, and inverse optimization, among others. All these mathematical models share an underlying common structure that allows us to conveniently formulate them in a unique format.

This book focuses on the analysis of this unifying format for equilibrium problems. Since it allows describing a large number of applications, many researchers devoted their efforts to study it, and nowadays, many results and algorithms are available: as optimization fits in this format, nonlinear programming techniques have often been the key tool of their work. The book aims at addressing in particular two core issues such as the existence and computation of equilibria. The first chapter illustrates a sample of applications, the second addresses the main theoretical issues, and the third introduces the main algorithms available for computing equilibria. The final chapter is devoted to quasi-equilibria, a more general format that is needed to cover more complex applications having additional features such as shared resources in noncooperative games. Finally, basic material on sets, functions, and multivalued maps that are exploited throughout the book are summarized in the appendix. To make the book as readable as possible, examples and applications have been included. We hope that this book may serve as a basis for a second-level academic course or a specialized course in a Ph.D. program and stimulate further interest in equilibrium problems.

Pisa, Italy   Giancarlo Bigi
L'Aquila, Italy   Marco Castellani
Pisa, Italy   Massimo Pappalardo
Pisa, Italy   Mauro Passacantando
July 2018

Contents

1 Equilibrium Models and Applications 1
   1.1   Obstacle Problem 1
   1.2   Power Control in Wireless Communications 3
   1.3   Traffic Network 4
   1.4   Portfolio Selection 6
   1.5   Optimal Production Under Restricted Resources 7
   1.6   Input-Output Analysis in an Economy 8
   1.7   Quality Control in Production Systems 10
   1.8   Ky Fan Inequalities: A Unifying Equilibrium Model 11
   1.9   Notes and References 16

2 Theory for Equilibria 17
   2.1   Convexity and Monotonicity 17
   2.2   Equivalent Reformulations 24
   2.3   Existence 31
   2.4   Stability 39
   2.5   Error Bounds 43
   2.6   Notes and References 47

3 Algorithms for Equilibria 51
   3.1   Fixed-Point and Extragradient Methods 51
   3.2   Descent Methods 59
   3.3   Regularization Methods 65
   3.4   Computational Issues 68
   3.5   Notes and References 69

4 Quasi-Equilibria 73
   4.1   Applications 73
   4.2   Theory 75
   4.3   Algorithms 80
   4.4   Notes and References 96

A **Mathematical Background** 99
   A.1 Topological Concepts 99
   A.2 Functions 103
   A.3 Multivalued Maps 108

References 113

Index 119

Chapter 1
Equilibrium Models and Applications

As already mentioned in the preface, the term "equilibrium" is widespread in science in the study of different phenomena. In this chapter a small selection of equilibrium problems from different areas is given, each leading to a different kind of mathematical model. The equilibrium position of an elastic string in presence of an obstacle, which is depicted in Sect. 1.1, coincides with the solution of a complementarity problem, the Nash equilibrium fits in well to model a power control multi-agent system described in Sect. 1.2, the steady distribution of traffic over a network is represented by a variational inequality in Sect. 1.3, the Markowitz portfolio theory is viewed as a multiobjective problem in Sect. 1.4, the shadow price theory is viewed as a saddle point problem for the nonlinear case in Sect. 1.5, the solution of the input-output model given in Sect. 1.6 is a fixed point and the quality control problem in a production system illustrated in Sect. 1.7 is an inverse optimization problem. Finally, the last section is devoted to show that all these mathematical models, which are apparently different, have a common structure that leads to a unified format: the Ky Fan inequality or the "equilibrium problem" using the "abstract" name introduced by Blum, Muu and Oettli to stress this unifying feature.

1.1 Obstacle Problem

Consider an elastic one-dimensional string bounded on a plane. The endpoints of the string are kept fixed, and its natural position is a straight line between the endpoints. What shape does the string take at the equilibrium if an obstacle is inserted in between the two endpoints? The string stretches over the obstacle: it sticks to the obstacle somewhere while it remains stretched as a straight line elsewhere (see Fig. 1.1).

© Springer Nature Switzerland AG 2019
G. Bigi et al., *Nonlinear Programming Techniques for Equilibria*, EURO Advanced
Tutorials on Operational Research, https://doi.org/10.1007/978-3-030-00205-3_1

Fig. 1.1 The obstacle problem

Mathematically, the equilibrium position of the string may be described by a function $u : [0, 1] \rightarrow \mathbb{R}$ with $u(0) = u(1) = 0$, with 0 and 1 denoting the x-coordinates of the two fixed endpoints. Analogously, the profile of the obstacle may be described by a function $f : [0, 1] \rightarrow \mathbb{R}$ with $f(0) < 0$ and $f(1) < 0$. At any point $x \in [0, 1]$ the string is above the obstacle, that is $u(x) \geq f(x)$. Moreover, the string and the obstacle touch, that is $u(x) = f(x)$, or the string is a straight line near x, that is $u''(x) = 0$. Therefore, if f has second order derivatives, then the shape of the string u has second order derivatives as well. Moreover, notice that the absence of any other force beyond the pressure of the obstacle guarantees that u is a concave function, which means $u''(x) \leq 0$ for any $x \in (0, 1)$. Summarising, the equilibrium position of the string is given by any u satisfying the conditions

$$\begin{cases} u(0) = u(1) = 0, \\ u(x) \geq f(x), \ u''(x) \leq 0 & x \in (0, 1), \\ (u(x) - f(x))u''(x) = 0 & x \in (0, 1). \end{cases} \qquad (1.1)$$

This functional system can be turned into a system of inequalities and equalities in finite dimension through piecewise linear approximations of u and f given on a finite grid of points. Specifically, fix any $n \in \mathbb{N}$ and consider $t_i = i/(n + 1)$, $u_i = u(t_i)$ and $f_i = f(t_i)$ for $i = 0, 1, \ldots, n + 1$. Exploiting finite differences, the second order derivative of u at t_i can be approximated through

$$u''(t_i) \approx (n + 1)^2 (u_{i+1} - 2u_i + u_{i-1}), \qquad i = 1, \ldots, n.$$

Therefore, system (1.1) can be approximated by the following system

$$\begin{cases} u_0 = u_{n+1} = 0, \\ u_i - f_i \geq 0, \ u_{i+1} - 2u_i + u_{i-1} \leq 0 & i = 1, \ldots, n, \\ (u_i - f_i)(u_{i+1} - 2u_i + u_{i-1}) = 0 & i = 1, \ldots, n, \end{cases} \qquad (1.2)$$

in the n variables $u_1, \ldots, u_n$. In turn, (1.2) can be written in a compact form introducing the $n \times n$ tridiagonal matrix

$$
M = \begin{bmatrix} 2 & -1 & & 0 \\ -1 & \ddots & \ddots & \\ & \ddots & \ddots & -1 \\ 0 & & -1 & 2 \end{bmatrix}
$$

and the vector $q \in \mathbb{R}^n$ with $q_i = 2f_i - f_{i+1} - f_{i-1}$. Indeed, $\bar{u} \in \mathbb{R}^n$ satisfies (1.2) if and only if $\bar{x} = \bar{u} - f$ satisfies

$$
\bar{x} \geq 0, \qquad M\bar{x} + q \geq 0, \qquad \langle \bar{x}, M\bar{x} + q \rangle = 0, \qquad (1.3)
$$

where the inequality $\geq$ is meant componentwise. Notice that the nonnegative conditions imply either $\bar{x}_i = 0$ or $(M\bar{x} + q)_i = 0$ for $\langle \bar{x}, M\bar{x} + q \rangle = 0$ to hold. Systems like (1.3) are known as (linear) *complementarity problems* since they require that the product of nonnegative quantities should be zero.

1.2 Power Control in Wireless Communications

A cellular network is designed to provide several users with access to wireless services over a large area that is divided into smaller areas called cells: each of them represents the area covered by a single base station which is often located at the center of the cell. In a code-division multiple-access (CDMA) system, mobile users operate using the same frequency and they need to adjust their transmit power to ensure a good performance (e.g., in terms of quality of service) by controlling the interference while minimizing the overall cost at the same time.

For the sake of simplicity, consider a single-cell CDMA system with N mobile users. Each user i has to select a value for the uplink transmit power $x_i \geq 0$ in order to minimize its own cost function

$$
c_i(x_i, x_{-i}) = \lambda_i x_i - \alpha_i \log(1 + \gamma_i(x)), \qquad (1.4)
$$

where $x_{-i} = (x_j)_{j \neq i}$ is the transmit power of all users except i and γ_i is the Signal-to-Interference-plus-Noise Ratio (SINR) function

$$
\gamma_i(x) = \frac{W}{R} \frac{h_i x_i}{\sum\limits_{j \neq i} h_j x_j + \sigma^2},
$$

where W is the chip rate, R is the total rate, $h_j \in (0, 1)$ is the channel gain from user j to the base station in the cell and σ^2 is the Gaussian noise variance. The objective function of each user is the difference between a pricing function (that assigns a price λ_i per power unit) and the gain obtained from a better SINR (α_i is called benefit parameter and represents the desired level of SINR). Therefore, an increase of the power level on one hand implies a benefit in terms of interference and on another hand a price in terms of the power consumed.

In this framework a vector of power levels chosen by the mobile users provides an equilibrium state if no user can decrease its own cost function by changing its power level unilaterally. More generally, a situation in which several selfish decision makers interact each other is known as *noncooperative game* and the equilibrium concept given above as *Nash equilibrium*.

1.3   Traffic Network

Consider a traffic network represented by a set of nodes N, a set of arcs $A \subseteq N \times N$ and a set OD of pairs of nodes that represent the origin and the destination of paths. For each pair $s \in OD$ there is a known demand d_s representing the rate of traffic entering and exiting the network at the origin and the destination of s respectively. The demand d_s has to be distributed among a given set P_s of paths connecting the pair s and let x_p denote the portion of d_s routed on path p. Let P be the set of all the n paths, i.e., the union of all the sets P_s over all $s \in OD$, and $x = (x_p)_{p \in P}$ the vector of all path flows. The set of feasible path flows is hence given by

$$C = \left\{ x \in \mathbb{R}_+^n : \quad \sum_{p \in P_s} x_p = d_s, \quad \forall s \in OD \right\}. \tag{1.5}$$

Since the flow z_a on each arc a is the sum of all flows on paths to which the arc belongs, the arc flow vector $z = (z_a)_{a \in A}$ is given by $z = \Delta x$, where Δ is the arc-path incidence matrix:

$$\Delta_{a,p} = \begin{cases} 1 \text{ if arc } a \in p, \\ 0 \text{ otherwise.} \end{cases}$$

A nonnegative arc cost function $t_a(z)$, which represents the travel time in traversing arc a and depends upon the whole arc flow vector z, is given for each arc a. Assuming that the path cost function is additive, the travel time $T_p(x)$ on path p is equal to the sum of the travel times on all the arcs of path p, that is

$$T_p(x) = \sum_{a \in p} t_a(\Delta x).$$

Therefore, the path cost map is $T : \mathbb{R}^n \to \mathbb{R}^n$ with $T(x) = \Delta^T t(\Delta x)$.

According to the *Wardrop equilibrium principle*, a vector $\bar{x} \in C$ is an equilibrium flow if it is positive only on paths with minimum cost, i.e., the following implication

$$\bar{x}_p > 0 \implies T_p(\bar{x}) = \min_{q \in P_s} T_q(\bar{x})$$

holds for any $s \in OD$ and $p \in P_s$.

It is possible to prove that a path flow $\bar{x} \in C$ is a Wardrop equilibrium if and only if the inequality

$$\langle T(\bar{x}), y - \bar{x} \rangle \geq 0, \qquad \forall \, y \in C, \tag{1.6}$$

holds. In fact, setting $\widetilde{T}_s = \min_{p \in P_s} T_p(\bar{x})$ for any $s \in OD$, if $\bar{x}$ is a Wardrop equilibrium, then any $y \in C$ satisfies

$$
\begin{aligned}
\langle T(\bar{x}), y - \bar{x} \rangle &= \sum_{s \in OD} \sum_{p \in P_s} T_p(\bar{x})(y_p - \bar{x}_p) \\
&= \sum_{s \in OD} \left[\sum_{p \in P_s: \, \bar{x}_p > 0} T_p(\bar{x})(y_p - \bar{x}_p) + \sum_{p \in P_s: \, \bar{x}_p = 0} T_p(\bar{x})(y_p - \bar{x}_p) \right] \\
&= \sum_{s \in OD} \left[\sum_{p \in P_s: \, \bar{x}_p > 0} \widetilde{T}_s(y_p - \bar{x}_p) + \sum_{p \in P_s: \, \bar{x}_p = 0} T_p(\bar{x}) y_p \right] \\
&\geq \sum_{s \in OD} \left[\sum_{p \in P_s: \, \bar{x}_p > 0} \widetilde{T}_s(y_p - \bar{x}_p) + \sum_{p \in P_s: \, \bar{x}_p = 0} \widetilde{T}_s y_p \right] \\
&= \sum_{s \in OD} \widetilde{T}_s \sum_{p \in P_s} (y_p - \bar{x}_p) \\
&= \sum_{s \in OD} \widetilde{T}_s (d_s - d_s) = 0,
\end{aligned}
$$

where the third equality follows from the definition of Wardrop equilibrium, the inequality from the definition of $\widetilde{T}_s$ and the fifth equality from the feasibility of y and $\bar{x}$. Thus, inequality (1.6) is satisfied. Conversely, if inequality (1.6) holds, then consider an arbitrary pair $s \in OD$, two paths $p, q \in P_s$ with $\bar{x}_p > 0$ and the path flow y defined as follows:

$$
y_r = \begin{cases} \bar{x}_r & \text{if } r \neq p, q, \\ 0 & \text{if } r = p, \\ \bar{x}_p + \bar{x}_q & \text{if } r = q. \end{cases}
$$

Then, it is clear that $y \in C$ and

$$0 \leq \langle T(\bar{x}), y - \bar{x} \rangle = T_p(\bar{x})(y_p - \bar{x}_p) + T_q(\bar{x})(y_q - \bar{x}_q) = \bar{x}_p(T_q(\bar{x}) - T_p(\bar{x})),$$

hence $T_q(\bar{x}) \geq T_p(\bar{x})$. Since path $q \in P_s$ is arbitrary, $T_p(\bar{x}) = \min_{q \in P_s} T_q(\bar{x})$ holds and hence $\bar{x}$ is a Wardrop equilibrium.

Inequalities like (1.6) are known as *variational inequalities*.

1.4 Portfolio Selection

Suppose there are n risky assets, where asset i gives the random return R_i. Recall that the return R of an asset is simply the percentage change in the value from one time to another; more precisely, the return at time t is defined by

$$R = \frac{V_t - V_{t-1}}{V_{t-1}},$$

where V_t is the total value of the asset at time t and V_{t-1} is the total value at an earlier time $t - 1$. For the sake of simplicity, assume that R_i are n jointly distributed random variables with finite second moment.

Let M be a given sum of money to be invested in the n different assets and let x_i denote the amount to be allocated to the asset i. The vector $x \in \mathbb{R}^n$ is called a *portfolio* if

$$x_1 + x_2 + \cdots + x_n \leq M.$$

Notice that the non-negativeness of the components x_i could not be required. A negative x_i represents a *short position* for the risky asset i: a short position is an investment strategy where the investor sells shares of borrowed stock in the open market. The expectation of the investor is that the price of the stock will decrease over time, at which point he will purchase the shares in the open market and return the shares to the broker which he borrowed them from.

Fixed a portfolio x, the net profit is described by the random variable

$$R = x_1 R_1 + x_2 R_2 + \cdots + x_n R_n.$$

A first simple method for establishing the goodness of the portfolio could be to maximize the expected return

$$g_1(x) = E[x_1 R_1 + x_2 R_2 + \ldots + x_n R_n] = x_1 \mu_1 + x_2 \mu_2 + \ldots + x_n \mu_n,$$

where $\mu_i = E[R_i]$, that is solving the following linear programming problem

$$\begin{cases} \max\ \mu_1 x_1 + \mu_2 x_2 + \ldots + \mu_n x_n \\ x_1 + x_2 + \ldots + x_n \leq M \\ x_i \geq 0, \quad i = 1, \ldots, n. \end{cases}$$

The constraints $x_i \geq 0$ mean that only long positions are allowed, hence these conditions should be omitted to include possible short sales in the model.

However, this approach does not provide meaningful results. Indeed, if there exists an index j such that $\mu_j > 0$ and $\mu_j > \mu_i$ for each $i \neq j$, then the problem has a unique optimal solution $\bar{x}$ with $\bar{x}_j = M$ and all the other components equal to zero, which represent an undiversified portfolio and it could be very risky: don't put all your eggs in one basket!

Risk aversion is the behaviour of investors, when exposed to uncertainty, to prefer a bargain with a more certain, but possibly lower, expected payoff rather than another bargain with an uncertain payoff. One simple measure of financial risk is the variance of the net profit, i.e.

$$g_2(x) = \text{Var}(x_1 R_1 + x_2 R_2 + \ldots + x_n R_n) = \sum_{i=1}^{n} \sum_{j=1}^{n} \sigma_{ij} x_i x_j,$$

where

$$\sigma_{ij} = \text{Cov}(R_i, R_j) = \text{E}[R_i R_j] - \text{E}[R_i]\text{E}[R_j]$$

is the covariance between R_i and R_j.

The investor would like to maximize the expected return g_1 and to minimize the risk measured by variance g_2. Since these two objectives are typically conflicting, more reasonable solutions are available by considering a trade-off between risk and return according to the Markowitz theory. One simple way to perform it is to consider portfolios such that no other portfolio provides a larger return paired with a lower risk. Formally, a portfolio x is *strictly dominated* by another portfolio x' if $g_1(x') > g_1(x)$ and $g_2(x') < g_2(x)$. Hence, the corresponding selection problem consists in finding a portfolio $\bar{x} \in \mathbb{R}_+^n$ which is not strictly dominated by any other portfolio. Such portfolios are called weak Pareto optimal.

1.5 Optimal Production Under Restricted Resources

A company produces a mix of n commodities aiming at maximizing the profit while also evaluating further investments in the m raw materials (or resources) that are needed. The profit is given by a function $f : \mathbb{R}^n \to \mathbb{R}$, where each variable x_j represents the quantity of commodity j that is produced. The use of each resource i depends on the total production $x = (x_1, \ldots, x_n)$ and is given by the function $g_i : \mathbb{R}^n \to \mathbb{R}$. Therefore, the maximization of the profit amounts to maximizing $f(x)$ subject to the constraints $g_i(x) \leq c_i$, where c_i denotes the available quantity of resource i.

The tool to evaluate further investments in resources are the so-called shadow prices, that is the marginal changes of the value of the optimal profit as the quantities

c_i vary. According to the theory of Lagrange multipliers, under a standard constraint qualification, if a mix $\bar{x}$ is optimal then there exists a nonnegative vector of shadow prices $\bar{\lambda}$ satisfying the following conditions:

$$\sum_{i=1}^{m} \bar{\lambda}_i \frac{\partial g_i}{\partial x_j}(\bar{x}) \geq \frac{\partial f}{\partial x_j}(\bar{x}), \qquad j = 1, \ldots, n, \qquad (1.7)$$

$$\bar{\lambda}_i (c_i - g_i(\bar{x})) = 0, \qquad i = 1, \ldots, m, \qquad (1.8)$$

$$\bar{x}_j \left[\frac{\partial f}{\partial x_j}(\bar{x}) - \sum_{i=1}^{m} \bar{\lambda}_i \frac{\partial g_i}{\partial x_j}(\bar{x}) \right] = 0, \qquad j = 1, \ldots, n. \qquad (1.9)$$

They yield the following economic interpretation: the left hand side of (1.7) is the marginal value of the amount of the resources needed to increase the production of commodity j while the right hand side is the marginal profit due to the increase; (1.8) means that if the resource i is not used completely then it must be free, i.e., a resource with zero shadow price, while if it is positive, then all of the available supply must be fully used; (1.9) asserts that if commodity j is produced, i.e., $\bar{x}_j > 0$, then the marginal quantities in (1.7) must be equal, while if they are not equal then commodity j cannot be produced.

If f is concave and g_i are convex, then the existence of such shadow prices $\bar{\lambda}$ is also a sufficient optimality condition for the mix $\bar{x}$. Moreover, the couple $(\bar{x}, \bar{\lambda})$ satisfies the conditions (1.7)–(1.9) if and only if it is a saddle point of the function

$$L(x, \lambda) = f(x) + \sum_{i=1}^{m} \lambda_i (c_i - g_i(x)), \qquad (1.10)$$

which is called the Lagrangian function; this means

$$L(x, \bar{\lambda}) \leq L(\bar{x}, \bar{\lambda}) \leq L(\bar{x}, \lambda), \quad \forall \, x, \lambda \geq 0,$$

or, equivalently, $\bar{x}$ maximizes $L(\cdot, \bar{\lambda})$ and $\bar{\lambda}$ minimizes $L(\bar{x}, \cdot)$.

1.6 Input-Output Analysis in an Economy

Input-output mathematical models for the economy of a country are based on its disaggregation into sectors. Suppose the economy consists of n interdependent sectors (or industries) $S_1, \ldots, S_n$ each of which produces a single kind of good that is traded, consumed and invested within the same economy. Let x_i denote the quantity of good i produced by the sector S_i. Each sector utilizes some of the goods produced by the other industries for the production of its own good. More precisely, suppose that the sector S_j must use y_{ij} units of the good i in order to produce

x_j units of the good j. The proportionality is the main assumption of the original Leontief input-output model: the exploitation of the good i is directly proportional to the production of the good j. In other words, the ratio $a_{ij} = y_{ij}/x_j$, called input coefficient, is constant and represents the units of good i needed for producing one unit of good j. Clearly, all the coefficients a_{ij} are nonnegative and moreover the possibility that $a_{ii} > 0$ holds for some i is not ruled out (for instance, a power station may use some of its own electric power for the production). The quantity

$$a_{i1}x_1 + a_{i2}x_2 + \cdots + a_{in}x_n$$

is the *internal demand* of the good i. In addition to the internal demand, which models the flow of goods in between the industries, suppose the existence of other non-productive sectors of the economy (such as consumers and governments), that may be grouped into the so-called *open sector* not producing anything but consuming goods from all the sectors. Denote by d_i the demand of the open sector from the sector S_i, which is called *final demand*. Therefore, the total output of the sector S_i must be equal to the internal plus the final demand:

$$x_i = a_{i1}x_1 + a_{i2}x_2 + \cdots + a_{in}x_n + d_i.$$

The output levels required by all the n sectors in order to meet these demands are given by the system of n linear equations coupled with nonnegativity conditions

$$\begin{cases} a_{11}x_1 + a_{12}x_2 + \ldots + a_{1n}x_n + d_1 = x_1 \\ a_{21}x_1 + a_{22}x_2 + \ldots + a_{2n}x_n + d_2 = x_2 \\ \vdots \\ a_{n1}x_1 + a_{n2}x_2 + \ldots + a_{nn}x_n + d_n = x_n \\ x_i \geq 0, \quad i = 1, \ldots, n \end{cases}$$

that can be written in matrix form as

$$\begin{cases} Ax + d = x \\ x \in \mathbb{R}_+^n \end{cases} \tag{1.11}$$

where the *input-output matrix* $A = (a_{ij})$ describes the interdependence of the industries.

The linearity assumption on the relation between each x_j and the amount y_{ij} is a very strong assumption. The assumption of constant returns to scale is arguable on the grounds that functions more complex than simple proportions are needed to describe production processes realistically, particularly in industries where at least one large installation (such as railroad tracks, dams or telephone lines) must be provided before any output can be produced. For this reason some authors proposed a nonlinear input-output model replacing the linear production functions $y_{ij} = a_{ij}x_j$

with the nonlinear functions $y_{ij} = a_{ij}(x_j)$. The basic assumptions on the nonlinear functions are the following:

- $a_{ij}(\cdot)$ is defined and continuously differentiable on $\mathbb{R}_+$,
- $a_{ij}(0) = 0$,
- $a'_{ij}(t) \geq 0$ for all $t \geq 0$.

Therefore, given the map $A : \mathbb{R}_+^n \to \mathbb{R}_+^n$ whose component i is defined by $A_i(x) = a_{i1}(x_1) + a_{i2}(x_2) + \ldots + a_{in}(x_n)$, the solution of the nonlinear input-output model consists in finding $\bar{x} \in \mathbb{R}_+^n$ such that $A(\bar{x}) + d = \bar{x}$, that is $\bar{x}$ is a fixed point of the map $x \mapsto A(x) + d$.

1.7 Quality Control in Production Systems

A manufacturer produces one commodity aiming at maximizing the profit while controlling the quality level of the production at its facility over a planning horizon of n time periods. Indeed, the quality level affects both the expected demand of the commodity and the cost of its production.

Let $x_i \in \mathbb{R}$ denote the quality level of the production during period i; it may be assumed $x_i \in [0, 1]$ without any loss of generality. Suppose $f_i : \mathbb{R}_+ \to \mathbb{R}_+$ is the expected demand and $g_i : \mathbb{R}_+ \to \mathbb{R}_+$ the production cost as a function of the quality level x_i during period i. If $D > 0$ denotes the total budget for producing the commodity and c_i the unitary price for period i, the problem can be modeled as the following mathematical program

$$\max \left\{ \sum_{i=1}^{n} c_i f_i(x_i) : \sum_{i=1}^{n} g_i(x_i) \leq D, \ \ell_i \leq x_i \leq u_i \right\} \tag{1.12}$$

for some given lower and upper bounds $0 \leq \ell_i \leq u_i \leq 1$ on the quality level. In some production systems $x_i \leq x_{i+1}$ (for $i = 1, \ldots, n-1$) are reasonable additional constraints: the quality level of the production does not get lower over time.

Actually, modifying the quality level may be a difficult task and may require a lot of time in some particular production systems. In such situations it may be convenient to adjust prices and it is clearly unreasonable to increase or decrease each price too much. Therefore, some $\bar{c} = (\bar{c}_1, \ldots, \bar{c}_n)$ sufficiently close to the current prices $c^* = (c_1^*, \ldots, c_n^*)$ is sought such that the current quality levels $\bar{x} = (\bar{x}_1, \ldots, \bar{x}_n)$ provide an optimal solution. The problem can be formally stated as follows:

given a feasible solution $\bar{x}$ of (1.12), $c^* \in \mathbb{R}_+^n$ and $\delta > 0$,

find $\bar{c} \in \mathbb{R}_+^n$ s.t. $\|\bar{c} - c^*\|_\infty \leq \delta$ and $\bar{x}$ is an optimal solution of (1.12).

Problems like the above one are called *inverse optimization problems* since they aim at determining whether a given feasible point can be made optimal by adjusting the values of some parameters within a given range.

1.8   Ky Fan Inequalities: A Unifying Equilibrium Model

In this section all the mathematical models of the problems described in the previous sections are recast as particular cases of a *Ky Fan inequality*, that is the following mathematical equilibrium model

$$\text{find } \bar{x} \in C \text{ such that } f(\bar{x}, y) \geq 0 \text{ for all } y \in C, \qquad \text{EP}(f, C)$$

where $C \subseteq \mathbb{R}^n$ is a nonempty closed set and $f : \mathbb{R}^n \times \mathbb{R}^n \rightarrow \mathbb{R}$ is an equilibrium bifunction, i.e., $f(x, x) = 0$ for all $x \in C$. Precisely, the section aims at showing how complementarity problems, Nash equilibrium problems, variational inequalities, weak Pareto optimization problems, saddle point problems, fixed point problems and inverse optimization problems can be all formulated in the above format through suitable choices of f and C.

Complementarity Problems

Given a closed convex cone $C \subseteq \mathbb{R}^n$ and a map $F : \mathbb{R}^n \rightarrow \mathbb{R}^n$, the complementarity problem asks to

$$\text{find } \bar{x} \in C \text{ such that } F(\bar{x}) \in C^* \text{ and } \langle \bar{x}, F(\bar{x}) \rangle = 0, \qquad (1.13)$$

where $C^* = \{y \in \mathbb{R}^n : \langle y, x \rangle \geq 0, \ \forall x \in C\}$ is the dual cone of C. Notice that the complementarity problem (1.3) described in Sect. 1.1 is a special case of (1.13) with $C = \mathbb{R}^n_+$ and $F(x) = Mx + q$.

Solving the complementarity problem amounts to solving EP(f, C) with

$$f(x, y) = \langle F(x), y - x \rangle.$$

Indeed, if $\bar{x}$ solves the complementarity problem, then

$$f(\bar{x}, y) = \langle F(\bar{x}), y - \bar{x} \rangle = \langle F(\bar{x}), y \rangle \geq 0, \qquad \forall y \in C,$$

that is $\bar{x}$ solves EP(f, C). Conversely, if $\bar{x}$ solves EP(f, C), then choosing $y = 2\bar{x}$ and $y = 0$ provides $\langle \bar{x}, F(\bar{x}) \rangle = 0$ and thus $\langle F(\bar{x}), y \rangle = f(\bar{x}, y) \geq 0$ holds for all $y \in C$, that is $\bar{x}$ is a solution of the complementarity problem.

Note that the system of equations $F(x) = 0$ is a special complementarity problem with $C = \mathbb{R}^n$.

Nash Equilibrium Problems

In a noncooperative game with N players, each player i has a set of possible strategies $C_i \subseteq \mathbb{R}^{n_i}$ and aims at minimizing a cost function $c_i : C \to \mathbb{R}$ with $C = C_1 \times \cdots \times C_N$. A Nash equilibrium is any $\bar{x} \in C$ such that no player can reduce its cost by unilaterally changing its strategy, that is any $\bar{x} \in C$ such that

$$c_i(\bar{x}_i, \bar{x}_{-i}) \leq c_i(y_i, \bar{x}_{-i})$$

holds for any $y_i \in C_i$ and any $i = 1, \ldots, N$, where $\bar{x}_{-i}$ denotes the vector of strategies of all players except i. Finding a Nash equilibrium amounts to solving $EP(f, C)$ with the so-called Nikaido-Isoda bifunction, i.e.,

$$f(x, y) = \sum_{i=1}^{N} [c_i(y_i, x_{-i}) - c_i(x_i, x_{-i})]. \tag{1.14}$$

Indeed, if $\bar{x}$ is a Nash equilibrium, all the terms in (1.14) are nonnegative for any $y \in C$ and hence $\bar{x}$ solves the equilibrium problem. Conversely, let $\bar{x}$ be a solution of $EP(f, C)$ and assume, by contradiction, there exist an index i and a strategy $y_i \in C_i$ such that $c_i(\bar{x}_i, \bar{x}_{-i}) > c_i(y_i, \bar{x}_{-i})$. Choosing $y_j = \bar{x}_j$ for all $j \neq i$ leads to the contradiction

$$f(\bar{x}, y) = c_i(y_i, \bar{x}_{-i}) - c_i(\bar{x}_i, \bar{x}_{-i}) < 0.$$

The power control game described in Sect. 1.2 is a Nash equilibrium problem where the strategy sets are $C_i = [0, +\infty)$ and the cost functions are defined in (1.4).

Variational Inequality Problems

Given a closed convex set $C \subseteq \mathbb{R}^n$ and a map $F : \mathbb{R}^n \to \mathbb{R}^n$, the variational inequality problem asks to

$$\text{find } \bar{x} \in C \text{ such that } \langle F(\bar{x}), y - \bar{x} \rangle \geq 0 \text{ for all } y \in C. \tag{1.15}$$

Solving this problem amounts to solving $EP(f, C)$ with

$$f(x, y) = \langle F(x), y - x \rangle.$$

Notice that the variational inequality (1.6) which models the traffic equilibrium problem in Sect. 1.3 is a special case of (1.15) with F equal to the path cost map T.

More general formats of variational inequalities are included in the $EP(f, C)$ format. For instance, if $F : \mathbb{R}^n \rightrightarrows \mathbb{R}^n$ is a multivalued map with compact values, then

$$\text{find } \bar{x} \in C \text{ and } \bar{u} \in F(\bar{x}) \text{ such that } \langle \bar{u}, y - \bar{x} \rangle \geq 0 \text{ for all } y \in C,$$

amounts to solving $EP(f, C)$ with

$$f(x, y) = \max_{u \in F(x)} \langle u, y - x \rangle.$$

Given two maps $F, G : \mathbb{R}^n \to \mathbb{R}^n$ and a function $h : \mathbb{R}^n \to (-\infty, +\infty]$, another kind of generalized variational inequality problem asks to

$$\text{find } \bar{x} \in \mathbb{R}^n \text{ such that } \langle F(\bar{x}), y - G(\bar{x}) \rangle + h(y) - h(G(\bar{x})) \geq 0 \text{ for all } y \in \mathbb{R}^n.$$

Solving this problem amounts to solving $EP(f, C)$ with $C = \mathbb{R}^n$ and

$$f(x, y) = \langle F(x), y - G(x) \rangle + h(y) - h(G(x)).$$

Notice that the presence of G and h does not allow formulating this problem in the standard format (1.15).

Weak Pareto Optimization Problems

Given m real-valued functions $\psi_i : \mathbb{R}^n \to \mathbb{R}$, $\bar{x} \in X$ is called a weak Pareto global minimum of the vector function $\psi = (\psi_1, \ldots, \psi_m)$ over a set $X \subseteq \mathbb{R}^n$ if there exists no element $y \in X$ such that $\psi_i(y) < \psi_i(\bar{x})$ for all $i = 1, \ldots, m$. Note that in the portfolio selection problem proposed in Sect. 1.4, the objective functions are $\psi_1(x) = -g_1(x)$ and $\psi_2(x) = g_2(x)$ with $X = \{x \in \mathbb{R}_+^n : x_1 + \cdots + x_n \leq M\}$.

Finding a weak Pareto global minimum amounts to solving $EP(f, C)$ with $C = X$ and

$$f(x, y) = \max_{i=1,\ldots,m} [\psi_i(y) - \psi_i(x)].$$

Indeed $f(\bar{x}, y) \geq 0$ for any $y \in X$ if and only if for any $y \in X$ there exists an index $i = 1, \ldots, m$ such that $\psi_i(y) - \psi_i(\bar{x}) \geq 0$, that is the definition of weak Pareto global minimum.

Saddle Point Problems

Given two sets $C_1 \subseteq \mathbb{R}^{n_1}$ and $C_2 \subseteq \mathbb{R}^{n_2}$, a saddle point of a function $H : C_1 \times C_2 \to \mathbb{R}$ is any $\bar{x} = (\bar{x}_1, \bar{x}_2) \in C_1 \times C_2$ such that

$$H(\bar{x}_1, y_2) \leq H(\bar{x}_1, \bar{x}_2) \leq H(y_1, \bar{x}_2)$$

holds for any $y = (y_1, y_2) \in C_1 \times C_2$. The production problem described in Sect. 1.5 is a saddle point problem where H is the opposite of the Lagrangian function (1.10).

Finding a saddle point of H amounts to solving EP(f, C) with $C = C_1 \times C_2$ and

$$f((x_1, x_2), (y_1, y_2)) = H(y_1, x_2) - H(x_1, y_2).$$

Indeed, a saddle point of H is a Nash equilibrium in a two-person zero-sum game, that is a noncooperative game where the cost function of the first player is H and the cost function of the second player is $-H$ (what one player wins is exactly what the other player loses).

Fixed Point Problems

Given a set $C \subseteq \mathbb{R}^n$, a fixed point of a map $F : C \to C$ is any $\bar{x} \in C$ such that $\bar{x} = F(\bar{x})$. For instance, the solution of the nonlinear input-output model described in Sect. 1.6 is a fixed point of the map $F(x) = A(x) + d$ over the set $C = \mathbb{R}_+^n$.

Finding a fixed point amounts to solving EP(f, C) with

$$f(x, y) = \langle x - F(x), y - x \rangle.$$

In fact, if $\bar{x}$ is a fixed point of F, then it obviously solves EP(f, C). Conversely, if $\bar{x}$ solves EP(f, C), then choosing $y = F(\bar{x}) \in C$ provides

$$0 \leq f(\bar{x}, F(\bar{x})) = -\|\bar{x} - F(\bar{x})\|^2,$$

hence $\bar{x} = F(\bar{x})$.

If the set C is also convex, a further equivalent reformulation is available, that is the fixed point problem amounts to solving EP(f, C) with

$$f(x, y) = \langle y - F(x), y - x \rangle.$$

In fact, if $\bar{x}$ is a fixed point of F, then $f(\bar{x}, y) = \|y - \bar{x}\|^2 \geq 0$ for any $y \in C$, i.e., $\bar{x}$ solves EP(f, C). Vice versa, if $\bar{x}$ is a solution to EP(f, C), then the convexity of C guarantees that $y = (\bar{x} + F(\bar{x}))/2 \in C$ and hence

$$0 \leq f(\bar{x}, y) = \left\langle \frac{\bar{x} - F(\bar{x})}{2}, \frac{F(\bar{x}) - \bar{x}}{2} \right\rangle = -\frac{1}{4}\|\bar{x} - F(\bar{x})\|^2,$$

thus $\bar{x} = F(\bar{x})$. Notice that the bifunctions defined in the two different reformulations satisfy different requirements: for instance, the former f is linear with respect to y while the latter is a strongly convex quadratic function in y.

Moreover, the EP(f, C) format includes also the fixed point problem when the map is multivalued. Indeed if $F : C \rightrightarrows C$ is a multivalued map with compact values, then finding $\bar{x} \in C$ such that $\bar{x} \in F(\bar{x})$ amounts to solving EP(f, C) with

$$f(x, y) = \max_{u \in F(x)} \langle x - u, y - x \rangle.$$

Inverse Optimization Problems

Given two closed sets $B \subseteq \mathbb{R}^n$ and $C \subseteq \mathbb{R}^m_+$, m functions $f_i : \mathbb{R}^n \rightarrow \mathbb{R}$ and p functions $h_j : \mathbb{R}^n \rightarrow \mathbb{R}$, the inverse optimization problem consists in determining a parameter $\bar{c} \in C$ such that at least one optimal solution of the maximization problem

$$\max \left\{ \sum_{i=1}^{m} \bar{c}_i f_i(x) \; : \; x \in B \right\}$$

satisfies the constraints $h_j(x) \le 0$ for all j. The problem of Sect. 1.7 can be formulated in this fashion where $m = n$, $p = 1$, B is the feasible region of (1.12), C is the set of vectors $c \in \mathbb{R}^n_+$ such that $\|c - c^*\|_\infty \le \delta$ and $h_1(x) = \|x - \bar{x}\|$.

Actually, this inverse optimization problem is equivalent to a noncooperative game with three players. The first player controls the variable $x \in B$ and aims at solving

$$\max \left\{ \sum_{i=1}^{m} c_i f_i(x) \; : \; x \in B \right\};$$

the second player controls the auxiliary variable $y \in \mathbb{R}^p_+$ and aims at solving

$$\max \left\{ \sum_{j=1}^{p} h_j(x) y_j \; : \; y \ge 0 \right\};$$

the third player controls the parameter $c \in C$ and aims at maximizing constant objective function, in other words he simply chooses the parameter c. Therefore, also this inverse optimization problem can be formulated in the EP(f, C) format via the Nikaido-Isoda bifunction (1.14).

1.9 Notes and References

Almost every paper on Ky Fan inequalities states that this model provides a general format that subsumes many other models so that a very large number of applications may be formulated in a unique fashion. The aim of this chapter is to corroborate this statement through a small selection of applications and the corresponding mathematical models. Section 1.1 examines the problem of determining the shape of an elastic string stretched over a body creating an obstacle [37]. This infinite dimensional problem can be numerically approximated by a complementarity system through standard discretization techniques. Section 1.2 describes the celebrated traffic network problem [110], which is based on the Wardrop equilibrium principle [128]. Its reformulation as a variational inequality was established in [54, 122]. Section 1.3 deals with a basic version of power control problems in wireless communications [69] that can be viewed as a noncooperative game and therefore turned into a system of inequalities through the Nikaido-Isoda aggregate bifunction [106]. Pareto optimization is the core of Sect. 1.4, that describes a simple portfolio selection problem relying on Markowitz's original approach [90]. Section 1.5 explains how optimal investment of resources, shadow prices and production of commodities are mixed in a mathematical model [18] that actually amounts to finding a saddle point of a suitable Lagrangian function. Section 1.6 describes the Leontief input-output model [89], which is probably the most well-known static model of the structure of a national economy. Indeed, Leontief received the Nobel Price in Economics "for the development of the input-output method and for its application to important economic problems" in 1973. The nonlinear version provided at the end of the section has been developed in [117]. The input-output analysis leads to a fixed point problem. Finally, inverse optimization (see [12]) allows modeling the problem of quality control in production systems that is addressed in Sect. 1.7 [133].

The solution set of each of the above problems coincides with the solution set of a Ky Fan inequality built by choosing a suitable feasible set C and a suitable bifunction f. This unified mathematical format has been explicitly proposed in [38, 99], following in the footsteps of the minimax inequality by Ky Fan [66].

Chapter 2
Theory for Equilibria

Basic theoretical topics such as the existence of solutions, their stability and error bounds are analysed for the Ky Fan inequality

$$\text{find } \bar{x} \in C \text{ such that } f(\bar{x}, y) \geq 0 \text{ for all } y \in C, \qquad \text{EP}(f, C)$$

where $C \subseteq \mathbb{R}^n$ is nonempty, convex and closed while $f : \mathbb{R}^n \times \mathbb{R}^n \to \mathbb{R}$ is an equilibrium bifunction, i.e., it satisfies $f(x, x) = 0$ for any $x \in \mathbb{R}^n$.

Existence results are achieved by appropriately pairing continuity, convexity and monotonicity assumptions on f, while suitable coercivity conditions may replace the compactness of the feasible region C. The same ingredients can be exploited to analyse the sensitivity of solutions to small changes in the data and to estimate the distance of a given point from the sets of the solutions. Tools and techniques for turning $\text{EP}(f, C)$ into another Ky Fan inequality, a suitable variational inequality or optimization problem with the same set of solutions are analysed as well. Some of these tools play an important role in the development of the algorithms of Chap. 3.

2.1 Convexity and Monotonicity

Convexity and monotonicity concepts that are useful throughout the book are collected in this section along with their main properties.

A set $C \subseteq \mathbb{R}^n$ is said to be *convex* if every point on the straight line that joins any pair of points of C lies also in C, i.e., $tx + (1 - t)y \in C$ for any $x, y \in C$ and $t \in [0, 1]$.

Real-valued functions are supposed to be defined on the whole space $\mathbb{R}^n$ (for instance, the function g means $g : \mathbb{R}^n \to \mathbb{R}$) throughout all this section.

© Springer Nature Switzerland AG 2019
G. Bigi et al., *Nonlinear Programming Techniques for Equilibria*, EURO Advanced
Tutorials on Operational Research, https://doi.org/10.1007/978-3-030-00205-3_2

A function g is *convex* on a convex set C if any $x, y \in C$ and $t \in [0, 1]$ satisfy

$$g(tx + (1 - t)y) \leq tg(x) + (1 - t)g(y).$$

A function g is called *concave* on C if $-g$ is convex on C. The simple statement that g is convex means that g is convex on the whole $\mathbb{R}^n$. The same understanding will be used for all the definitions of the section that involve C.

Convexity preserving operations produce new convex functions out of a set of functions that are already known to be convex. A few simple convexity-preserving rules are given below.

– If g is convex on C and $\alpha \geq 0$, then αg is convex on C;
– if g_1, g_2 are convex on C, then $g_1 + g_2$ is convex on C;
– if g_i are convex on C for any $i \in I$, where I is an arbitrary set of indices such that $g(x) = \sup_{i \in I} g_i(x)$ is finite for any $x \in \mathbb{R}^n$, then g is convex on C.

Convex functions have a lot of properties. First, any convex function on C is continuous on the relative interior of C (which is always nonempty). Moreover, if g is convex on C, the sublevel sets

$$\{x \in C : g(x) < a\} \quad \text{and} \quad \{x \in C : g(x) \leq a\} \tag{2.1}$$

are convex sets for any $a \in \mathbb{R}$. Nevertheless, a function whose sublevel sets are convex sets may fail to be convex.

Example 2.1.1 Let $n = 1$. The function $g(x) = -e^x$ is not convex but the sublevel sets

$$\{x \in \mathbb{R} : g(x) \leq a\} = \begin{cases} \mathbb{R} & \text{if } a \geq 0 \\ [\ln(-a), +\infty) & \text{if } a < 0 \end{cases}$$

are convex for any $a \in \mathbb{R}$.

The family of functions with convex sublevel sets is worth consideration as well.

Definition 2.1.2 A function g is *quasiconvex* on C if the sublevel sets (2.1) are convex for each $a \in \mathbb{R}$. A function g is called *quasiconcave* on C if $-g$ is quasiconvex on C.

Quasiconvexity can be alternatively defined through suitable inequalities. Indeed, g is quasiconvex on C if and only if the inequality

$$g(tx + (1 - t)y) \leq \max\{g(x), g(y)\}$$

holds for any $x, y \in C$ and $t \in [0, 1]$.

When g is continuously differentiable, then it is convex on C if and only if one of the following equivalent inequalities

$$g(y) \geq g(x) + \langle \nabla g(x), y - x \rangle, \tag{2.2}$$

$$\langle \nabla g(x) - \nabla g(y), x - y \rangle \geq 0 \tag{2.3}$$

hold for any $x, y \in C$. Characterization (2.2) states that a function is convex on C if and only if the graph of the function lies above the tangent hyperplane to the graph at each $x \in C$. A map $F : \mathbb{R}^n \to \mathbb{R}^n$ is said to be *monotone* on C if

$$\langle F(x) - F(y), x - y \rangle \geq 0$$

for any $x, y \in C$. Therefore, characterization (2.3) states that a function is convex if and only if its gradient is a monotone map. It is worth noticing that when F is a continuously differentiable map, then it is monotone if and only if the Jacobian $\nabla F(x)$ is positive semidefinite at any x, that is

$$\langle z, \nabla F(x)z \rangle \geq 0, \qquad \forall x, z \in \mathbb{R}^n.$$

As a consequence, the twice continuously differentiable function g is convex if and only if its Hessian $\nabla^2 g(x)$ is positive semidefinite at any $x \in \mathbb{R}^n$.

A vector $x^* \in \mathbb{R}^n$ is said to be a *subgradient* of g at $x \in \mathbb{R}^n$ if the inequality

$$g(y) \geq g(x) + \langle x^*, y - x \rangle$$

holds for any $y \in \mathbb{R}^n$. The set of all the subgradients

$$\partial g(x) = \{x^* \in \mathbb{R}^n : g(y) \geq g(x) + \langle x^*, y - x \rangle \quad \forall y \in \mathbb{R}^n\}$$

is called the *subdifferential* of g at x. Since the domain of g is the whole space $\mathbb{R}^n$, then g is convex if and only if the subdifferential is a nonempty compact and convex set at any $x \in \mathbb{R}^n$. Moreover, a convex function g is differentiable at x if and only if $\partial g(x) = \{\nabla g(x)\}$. If g_1, g_2 are two convex functions then the subdifferential of $g_1 + g_2$ is the sum of the subdifferentials of g_1 and g_2

$$\partial(g_1 + g_2)(x) = \partial g_1(x) + \partial g_2(x). \tag{2.4}$$

Optimality for the minimization problem

$$\min\{g(x) : x \in C\} \tag{2.5}$$

can be fully characterized through subdifferentials whenever g is convex.

Theorem 2.1.3 *Suppose g is convex. Then,*

a) *any local minimum of (2.5) is also a global minimum;*
b) *the set of optimal solutions of (2.5) is a (possibly empty) convex set;*
c) *$\bar{x}$ solves (2.5) if and only if it is a stationary point, that is there exists $\bar{x}^* \in \partial g(\bar{x})$ such that the inequality*

$$\langle \bar{x}^*, y - \bar{x} \rangle \geq 0$$

holds for any $y \in C$.

Theorem 2.1.3 allows characterizing the projection $P_C(x)$ of a point x on a closed and convex set C. The projection is defined as the unique optimal solution of the minimization problem

$$\min\{\|y - x\|^2 : y \in C\}, \tag{2.6}$$

where the uniqueness of the optimal solution is guaranteed by the strong convexity of the objective function (see Definition 2.1.5 and Theorem 2.1.9).

Theorem 2.1.4

a) *$z = P_C(x)$ if and only if the inequality $\langle y - z, x - z \rangle \leq 0$ holds for any $y \in C$;*
b) *the map $P_C : \mathbb{R}^n \to \mathbb{R}^n$ is non-expansive, that is the inequality*

$$\|P_C(x) - P_C(y)\| \leq \|x - y\|$$

holds for any $x, y \in \mathbb{R}^n$.

A more general concept of convexity can be introduced in the following way.

Definition 2.1.5 Given $\tau \in \mathbb{R}$, a function g is called τ-*convex* on C if any $x, y \in C$ and $t \in [0, 1]$ satisfy

$$g(tx + (1 - t)y) \leq tg(x) + (1 - t)g(y) - \frac{\tau}{2}t(1 - t)\|x - y\|^2. \tag{2.7}$$

A function g is called τ-*concave* on C if $-g$ is τ-convex on C.

If $\tau = 0$, the above inequality provides the usual definition of convexity. If $\tau > 0$, g is also called *strongly convex* while it is called *weakly convex* if $\tau < 0$.

The following result provides an interesting characterization of τ-convexity.

Theorem 2.1.6 *The function g is τ-convex on C if and only if $g - \tau\|\cdot\|^2/2$ is convex on C.*

Proof Given any $x, y \in C$ and $t \in [0, 1]$, the equalities

$$\|tx + (1 - t)y\|^2 + t(1 - t)\|x - y\|^2 = t^2\|x\|^2 + (1 - t)^2\|y\|^2 + 2t(1 - t)\langle x, y\rangle$$
$$+ t(1 - t)(\|x\|^2 + \|y\|^2 - 2\langle x, y\rangle)$$
$$= t\|x\|^2 + (1 - t)\|y\|^2$$

follow immediately from the definition of the Euclidean norm. Thus, $g - \tau\|\cdot\|^2/2$ is convex if and only if

$$g(tx + (1 - t)y) - \tau\|tx + (1 - t)y\|^2/2 \le tg(x) - \tau t\|x\|^2/2$$
$$+ (1 - t)g(y) - \tau(1 - t)\|y\|^2/2$$
$$= tg(x) + (1 - t)g(y) - \tau t(1 - t)\|x - y\|^2/2$$
$$- \tau\|tx + (1 - t)y\|^2/2$$

holds for any $x, y \in C$ and $t \in [0, 1]$, that is g is τ-convex. $\qquad\square$

In particular the squared norm $\|\cdot\|^2$ is a τ-convex function with $\tau = 2$. Moreover, Theorem 2.1.6 and convexity preserving operations allow deducing the following properties.

Corollary 2.1.7

a) If g is τ-convex on C and $\alpha > 0$, then αg is $(\alpha\tau)$-convex on C;
b) if g_i are τ_i-convex on C with $i = 1, 2$ then $g_1 + g_2$ is $(\tau_1 + \tau_2)$-convex on C;
c) if g_i are τ-convex on C for any $i \in I$, where I is an arbitrary set of indices such that $g(x) = \sup_{i \in I} g_i(x)$ is finite for every $x \in \mathbb{R}^n$, then g is τ-convex on C.

Clearly, τ-convexity with $\tau \ge 0$ implies convexity and hence quasiconvexity as well. On the contrary, there is no relationships between τ-convexity and quasiconvexity when $\tau < 0$ as the following example shows.

Example 2.1.8 Let $n = 1$. The function $g(x) = -x^2/2$ is τ-convex with $\tau = -1$ but it is not quasiconvex. On the other hand, $g(x) = x^3$ is quasiconvex since it is increasing, but it is not τ-convex since $x^3 - \tau x^2/2$ is not convex on $(-\infty, \tau/6)$.

Theorem 2.1.9 *If g is τ-convex with $\tau > 0$, then the minimization problem (2.5) has a unique solution $\bar{x}$ and any $x \in C$ satisfies*

$$g(x) \ge g(\bar{x}) + \tau\|x - \bar{x}\|^2/2. \tag{2.8}$$

Proof The function $g - \tau\|\cdot\|^2/2$ is convex by Theorem 2.1.6. Choosing $x = 0$ and $x^* \in \partial\left(g - \tau\|\cdot\|^2/2\right)(0)$, inequality (2.2) provides

$$g(y) = \left(g(y) - \tau\|y\|^2/2\right) + \tau\|y\|^2/2 \ge g(0) + \langle x^*, y\rangle + \tau\|y\|^2/2 \to +\infty$$

as $\|y\| \to +\infty$. Since g is continuous, Theorem A.2.2 guarantees that (2.5) has at least one solution $\bar{x}$. Suppose that $\hat{x} \neq \bar{x}$ is another solution of (2.5). Then, the contradiction

$$g(t\hat{x} + (1-t)\bar{x}) \leq tg(\hat{x}) + (1-t)g(\bar{x}) - \tau t(1-t)\|\hat{x} - \bar{x}\|^2/2 < g(\bar{x})$$

arises for any $t \in (0, 1)$. Hence, the solution $\bar{x}$ is unique.

Moreover, the convexity of $g - \tau \| \cdot \|^2/2$ guarantees that the inequality

$$g(x) - \tau \|x\|^2/2 \geq g(\bar{x}) - \tau \|\bar{x}\|^2/2 + \langle x^* - \tau\bar{x}, x - \bar{x} \rangle$$

and equivalently

$$g(x) \geq g(\bar{x}) + \langle x^*, x - \bar{x} \rangle + \tau(\|x\|^2 + \|\bar{x}\|^2 - 2\langle \bar{x}, x \rangle)/2$$
$$= g(\bar{x}) + \langle x^*, x - \bar{x} \rangle + \tau \|x - \bar{x}\|^2/2$$

hold for any $x \in C$ and any $x^* \in \partial g(\bar{x})$. The optimality of $\bar{x}$ guarantees that $\langle \bar{x}^*, x - \bar{x} \rangle \geq 0$ holds for some $\bar{x}^* \in \partial g(\bar{x})$ so that (2.8) readily follows. $\square$

More general concepts of monotonicity can be introduced in the following way.

Definition 2.1.10 Given $\mu \in \mathbb{R}$, a map $F : \mathbb{R}^n \to \mathbb{R}^n$ is called

– μ-*monotone* on C if the inequality

$$\langle F(x) - F(y), x - y \rangle \geq \mu \|x - y\|^2 \tag{2.9}$$

holds for any $x, y \in C$;
– μ-*pseudomonotone* on C if the implication

$$\langle F(y), x - y \rangle \geq 0 \implies \langle F(x), x - y \rangle \geq \mu \|x - y\|^2$$

holds for any $x, y \in C$.

If $\mu = 0$, the above definitions collapse to the usual definitions of monotone and pseudomonotone maps. If $\mu > 0$, F is also called *strongly (pseudo)monotone* and similarly it is called *weakly (pseudo)monotone* if $\mu < 0$.

Exploiting the same kind of procedure for characterizing τ-convexity in Theorem 2.1.6, it is easy to show the μ-monotonicity of F is equivalent to the monotonicity of the map $x \mapsto F(x) - \mu x$. Therefore, when F is continuously differentiable, then it is μ-monotone if and only if $\nabla F - \mu I$ is positive semidefinite, where I denotes the identity matrix.

Moreover, any μ-monotone map is obviously also μ-pseudomonotone while the vice versa is not true.

Example 2.1.11 Let $n = 1$. The map $F(x) = x^2 + 1$ is pseudomonotone on $\mathbb{R}$. Indeed, the equivalences

$$F(y)(x - y) \geq 0 \iff x - y \geq 0 \iff F(x)(x - y) \geq 0$$

hold since F is positive on $\mathbb{R}$. On the contrary, F is not μ-monotone since, given any $\mu \in \mathbb{R}$, the inequality

$$[F(\bar{x}) - F(\bar{y})](\bar{x} - \bar{y}) = \mu - 1 < \mu = \mu(\bar{x} - \bar{y})^2$$

holds for $\bar{x} = \mu/2$ and $\bar{y} = -1 + \mu/2$.

The relationship between τ-convexity and μ-monotonicity for continuously differentiable functions follows immediately from Theorem 2.1.6 and characterization (2.3) of convex functions.

Corollary 2.1.12 *A continuously differentiable function g is τ-convex on C if and only if ∇g is τ-monotone on C.*

When the Ky Fan inequality EP(f, C) is a variational inequality, i.e., $f(x, y) = \langle F(x), y - x \rangle$ for some map $F : \mathbb{R}^n \to \mathbb{R}^n$, the left-hand side of (2.9) reads

$$\langle F(x) - F(y), x - y \rangle = -(f(x, y) + f(y, x)).$$

This relationship suggests a way to introduce monotonicity conditions for bifunctions that pairs those given in Definition 2.1.10 for maps.

Definition 2.1.13 Given $\mu \in \mathbb{R}$, a bifunction $f : \mathbb{R}^n \times \mathbb{R}^n \to \mathbb{R}$ is called

- μ-*monotone* on C if the inequality

$$f(x, y) + f(y, x) \leq -\mu \|x - y\|^2$$

holds for any $x, y \in C$;
- μ-*pseudomonotone* on C if the implication

$$f(x, y) \geq 0 \implies f(y, x) \leq -\mu \|y - x\|^2$$

holds for any $x, y \in C$.

If $\mu = 0$, f is also called (pseudo)monotone, if $\mu > 0$ *strongly* (pseudo)monotone while *weakly* (pseudo)monotone if $\mu < 0$. Clearly, μ-monotonicity implies μ-pseudomonotonicity but the converse does not hold as the following example shows.

Example 2.1.14 Let $n = 1$. The bifunction $f(x, y) = (2x^2 + xy + y^2 + 1)(x - y)$ is pseudomonotone on $\mathbb{R}$. Indeed, the equivalences

$$f(x, y) \geq 0 \iff x - y \geq 0 \iff f(y, x) = (2y^2 + xy + x^2 + 1)(y - x) \leq 0$$

hold since $2x^2 + xy + y^2 \geq 0$ is true on the whole $\mathbb{R}^2$. On the contrary, f is not μ-monotone since, given any $\mu \in \mathbb{R}$, the inequality

$$f(\bar{x}, \bar{y}) + f(\bar{y}, \bar{x}) = 1 - \mu > -\mu = -\mu(\bar{x} - \bar{y})^2$$

holds for $\bar{x} = 1 - \mu/2$ and $\bar{y} = -\mu/2$.

2.2 Equivalent Reformulations

Chapter 1 shows that different mathematical models have an underlying common structure that leads to the Ky Fan inequality $\mathrm{EP}(f, C)$. This section aims at tackling an issue that can be somehow considered the opposite procedure. Specifically, possible ways to reformulate the Ky Fan inequality as another problem with the same set of solutions are analysed.

Probably, the most straightforward reformulation relies on fixed points. Indeed, in order to turn $\mathrm{EP}(f, C)$ into a fixed point problem, consider the multivalued map $Y : \mathbb{R}^n \rightrightarrows \mathbb{R}^n$ given by

$$Y(x) = \arg\min\{f(x, y) : y \in C\},$$

which could be possibly empty at some points. The fixed points of Y coincide with the solutions of $\mathrm{EP}(f, C)$.

Theorem 2.2.1 *The point $\bar{x} \in C$ solves $\mathrm{EP}(f, C)$ if and only if $\bar{x} \in Y(\bar{x})$.*

Proof $\bar{x} \in Y(\bar{x})$ reads $0 = f(\bar{x}, \bar{x}) \leq f(\bar{x}, y)$ for all $y \in C$.   □

If $Y(x) \neq \emptyset$ for any $x \in \mathbb{R}^n$, then the value function

$$\varphi(x) = \max\{-f(x, y) : y \in C\} = -\min\{f(x, y) : y \in C\} \tag{2.10}$$

allows reformulating $\mathrm{EP}(f, C)$ as a constrained optimization problem.

Theorem 2.2.2 *Suppose $Y(x) \neq \emptyset$ for any $x \in \mathbb{R}^n$. Then, the following statements hold:*

a) $\varphi(x) \geq 0$ *for any $x \in C$;*
b) $\bar{x} \in \mathbb{R}^n$ *solves $\mathrm{EP}(f, C)$ if and only if $\bar{x} \in C$ and $\varphi(\bar{x}) = 0$.*

Proof If $x \in C$ and $y \in Y(x)$, then $\varphi(x) = -f(x, y) \geq -f(x, x) = 0$. Moreover, $\bar{x} \in C$ and $\varphi(\bar{x}) = 0$ if and only if $f(\bar{x}, y) \geq 0$ for all $y \in C$, i.e., it solves $\mathrm{EP}(f, C)$.   □

The function φ is called *gap function* and it is the simplest and most common merit function that can be associated to Ky Fan inequalities.

Definition 2.2.3 Given a set $X \subseteq \mathbb{R}^n$, a function $\psi : \mathbb{R}^n \to \mathbb{R}$ is said to be a *merit function on X for EP(f, C)* if

i) $\psi(x) \geq 0$ for any $x \in X$,
ii) $\bar{x} \in \mathbb{R}^n$ solves EP(f, C) if and only if $\bar{x} \in X$ and $\psi(\bar{x}) = 0$.

In the above definition X needs not to be necessarily the feasible region C of the Ky Fan inequality. Anyhow, notice that each solution of EP(f, C) solves the minimization problem

$$\min\{\psi(x) : x \in X\}. \tag{2.11}$$

The following example shows that solving EP(f, C) is not equivalent to solving (2.11). Indeed, the equivalence holds provided that the optimal value is zero.

Example 2.2.4 Consider EP(f, C) with the bifunction $f(x, y) = -\|x - y\|$ and the feasible region $C = \{x \in \mathbb{R}^n : \|x\| \leq 1\}$. Exploiting the triangle inequality property for the Euclidean norm, Y turns out to be

$$Y(x) = \begin{cases} \{y \in \mathbb{R}^n : \|y\| = 1\} & \text{if } x = 0 \\ -x/\|x\| & \text{if } x \neq 0 \end{cases}$$

and hence $\varphi(x) = 1 + \|x\|$, which is non-negative everywhere and it achieves its minimum at $\bar{x} = 0$. But $\bar{x}$ does not solve EP(f, C) since $f(\bar{x}, y) < 0$ holds for any $y \neq 0$. Indeed, $\bar{x}$ minimizes φ over $\mathbb{R}^n$ but $\varphi(0) = 1 \neq 0$.

Since each Ky Fan inequality can be formulated as a global optimization problem, minimization algorithms may be used for solving the optimization problem (2.11). The convexity of a merit function would make optimization easier since local minima are global minima and first-order necessary conditions are also sufficient conditions for optimality (Theorem 2.1.3).

Theorem 2.2.5 *Suppose $Y(x) \neq \emptyset$ for any $x \in \mathbb{R}^n$.*

a) *If $f(\cdot, y)$ is τ-concave for any $y \in \mathbb{R}^n$, then φ is τ-convex.*
b) *If $f(\cdot, y)$ is upper semicontinuous for any $y \in \mathbb{R}^n$, then φ is lower semicontinuous.*

Proof Since φ is the pointwise maximum of the functions $-f(\cdot, y)$, which are either τ-convex or lower semicontinuous, then Corollary 2.1.7 c) guarantees a) while b) is granted as well (see Appendix A.2). $\qquad\qquad\qquad\qquad\qquad\qquad\qquad\qquad\qquad\square$

Unfortunately, the gap function φ inherits all the troubles of Y: it might not be defined everywhere and it is not necessarily continuous on its domain. Nevertheless, the continuity of f and the compactness of C guarantee the continuity of φ.

Theorem 2.2.6 *Suppose C is compact and f is continuous. Then, Y is a nonempty upper semicontinuous multivalued map and the gap function φ is continuous. If in addition $Y(x) = \{y(x)\}$ is a singleton for any $x \in \mathbb{R}^n$, then the function $y(\cdot)$ is continuous.*

Proof The compactness of C and the continuity of f imply that $Y(x)$ is a nonempty and closed set for any $x \in \mathbb{R}^n$ by Weierstrass Theorem A.2.1. Consider any $\{(x^k, y^k)\}$ such that $y^k \in Y(x^k)$ and $(x^k, y^k) \to (x, y)$. The continuity of f guarantees that

$$f(x, y) = \lim_{k \to +\infty} f(x^k, y^k) \le \lim_{k \to +\infty} f(x^k, z) = f(x, z)$$

holds for any $z \in C$. Hence, $y \in Y(x)$ and Y is closed. Since $Y(\mathbb{R}^n) \subseteq C$ is bounded, the closedness of Y implies that it is upper semicontinuous by the Closed Graph Theorem A.3.4. Whenever $Y(x)$ is a singleton for any $x \in \mathbb{R}^n$, the continuity of $y(\cdot)$ readily follows.

Consider any $x \in \mathbb{R}^n$, any sequence $x^k \to x$ and any $y^k \in Y(x^k)$. A suitable subsequence $\{x^{k_\ell}\}$ satisfies

$$\lim_{\ell \to +\infty} \varphi(x^{k_\ell}) = \limsup_{k \to +\infty} \varphi(x^k).$$

Taking, if necessary, a further subsequence, $y^{k_\ell} \to y$ as $\ell \to +\infty$ holds for some $y \in C$ thanks to the compactness of C. Moreover, $y \in Y(x)$ since Y is closed. The chain of equalities

$$\limsup_{k \to +\infty} \varphi(x^k) = \lim_{\ell \to +\infty} \varphi(x^{k_\ell}) = \lim_{\ell \to +\infty} -f(x^{k_\ell}, y^{k_\ell}) = -f(x, y) = \varphi(x)$$

shows the upper semicontinuity of φ. Therefore, φ is continuous since it is lower semicontinuous by Theorem 2.2.5 *b*).   □

If f enjoys suitable differentiability properties, then φ is differentiable as well.

Theorem 2.2.7 *Suppose C is compact, f is continuous and $f(\cdot, y)$ is differentiable for any $y \in \mathbb{R}^n$ with $\nabla_x f$ being continuous (in both variables). Then, the one-sided directional derivative $\varphi'(x; d)$ of the gap function φ exists at any $x \in \mathbb{R}^n$ and any direction $d \in \mathbb{R}^n$ with*

$$\varphi'(x; d) = -\min\{\langle \nabla_x f(x, y), d \rangle : y \in Y(x)\}.$$

If in addition $Y(x) = \{y(x)\}$ is a singleton for any $x \in C$, then φ is continuously differentiable with

$$\nabla \varphi(x) = -\nabla_x f(x, y(x)).$$

Proof The multivalued map Y is nonempty and closed (see the proof of Theorem 2.2.6). Given any $x \in C$ and any direction $d \in \mathbb{R}^n$, the inequality

$$\liminf_{t \to 0^+} \frac{\varphi(x + td) - \varphi(x)}{t} \geq \liminf_{t \to 0^+} -\frac{f(x + td, y) - f(x, y)}{t} = -\langle \nabla_x f(x, y), d \rangle$$

holds for any fixed $y \in Y(x)$. The continuity of $\nabla_x f$ guarantees

$$\liminf_{t \to 0^+} \frac{\varphi(x + td) - \varphi(x)}{t} \geq -\min\{\langle \nabla_x f(x, y), d \rangle : y \in Y(x)\}. \tag{2.12}$$

Similarly, consider any $\{t_k\} \to 0^+$ and any $y^k \in Y(x + t_k d)$. The Mean Value Theorem A.2.6 guarantees the existence of $s_k \in [0, 1]$ such that

$$\frac{\varphi(x + t_k d) - \varphi(x)}{t_k} \leq \frac{-f(x + t_k d, y^k) + f(x, y^k)}{t_k} = -\langle \nabla_x f(x + s_k t_k d, y^k), d \rangle.$$

Since $x + t_k d \to x$, then $x + s_k t_k d \to x$ holds as well and the sequence $\{y^k\}$ is bounded. Thanks to the compactness of C and the closedness of Y, every cluster point of $\{y^k\}$ belongs to $Y(x)$. Consequently, the above inequality leads to

$$\limsup_{t \to 0^+} \frac{\varphi(x + td) - \varphi(x)}{t} \leq -\min\{\langle \nabla_x f(x, y), d \rangle : y \in Y(x)\}. \tag{2.13}$$

Equations (2.12) and (2.13) together guarantee that φ is directionally differentiable. The continuous differentiability is an immediate consequence of the continuity of $y(\cdot)$, which is guaranteed by Theorem 2.2.6.   □

Theorem 2.2.7 clearly shows the reason why φ is not necessarily differentiable even if f is sufficiently regular: Y has to be single-valued. Taking this into account, it is possible to investigate how to turn the Ky Fan inequality EP(f, C) into another with additional nice properties that enable to prove the existence of solutions or to devise solution methods.

Theorem 2.2.8 *Let* f_1, f_2 *be two equilibrium bifunctions such that* $f_1(x, \cdot)$ *and* $f_2(x, \cdot)$ *are convex for any* $x \in C$. *If*

$$\bigcup_{r>0} r \partial_y f_1(x, x) = \bigcup_{s>0} s \partial_y f_2(x, x) \tag{2.14}$$

holds for any $x \in C$, *then* EP(f_1, C) *and* EP(f_2, C) *have the same set of solutions.*

Proof If $\bar{x}$ solves EP(f_1, C), then it minimizes $f_1(\bar{x}, \cdot)$ over C, hence Theorem 2.1.3 c) guarantees the existence of $x_1^* \in \partial_y f_1(\bar{x}, \bar{x})$ such that $\langle x_1^*, y - \bar{x} \rangle \geq 0$ holds for any $y \in C$. Therefore, (2.14) guarantees the existence of $t > 0$ and $x_2^* \in \partial_y f_2(\bar{x}, \bar{x})$ such that $x_1^* = t x_2^*$ and thus $\bar{x}$ minimizes $f_2(\bar{x}, \cdot)$ over C by Theorem 2.1.3 c).

Hence, $f_2(\bar{x}, y) \geq f_2(\bar{x}, \bar{x}) = 0$ holds for any $y \in C$ and $\bar{x}$ solves EP(f_2, C). Exchanging the roles of f_1 and f_2, the thesis readily follows.   □

Two Ky Fan inequalities EP(f_1, C) and EP(f_2, C) may have the same set of solutions even though (2.14) does not holds.

Example 2.2.9 Let $n = 2$. The differentiable bifunctions $f_1(x, y) = y_1^2 + 2y_2^2 - x_1^2 - 2x_2^2$ and $f_2(x, y) = 2y_1^2 + y_2^2 - 2x_1^2 - x_2^2$ are convex with respect to y and the Ky Fan inequalities EP(f_1, C) and EP(f_2, C) have the same unique solution $(\bar{x}_1, \bar{x}_2) = (0, 0)$ for $C = [-1, 1] \times [-1, 1]$. Nevertheless, there does not exist any $\alpha > 0$ such that $\nabla_y f_2(x, x) = \alpha \nabla_y f_1(x, x)$, i.e., $\alpha(2x_1, 4x_2) = (4x_1, 2x_2)$, for any $(x_1, x_2) \in C$ with $x_1 x_2 \neq 0$.

Theorem 2.2.8 allows deducing some particular equivalent reformulations of EP(f, C) when $f(x, \cdot)$ is convex. Suppose $f(x, \cdot)$ is differentiable for any $x \in C$. The equilibrium bifunction $\hat{f}(x, y) = \langle \nabla_y f(x, x), y - x \rangle$ is linear with respect to y and $\partial_y \hat{f}(x, x) = \{\nabla_y f(x, x)\}$. Therefore, EP($f, C$) is equivalent to the variational inequality

$$\text{find } \bar{x} \in C \text{ such that } \langle \nabla_y f(\bar{x}, \bar{x}), y - \bar{x} \rangle \geq 0 \text{ for all } y \in C. \qquad (2.15)$$

Therefore, any method for variational inequalities could be applied to solve Ky Fan inequalities through (2.15). Clearly, all the assumptions on the operator of the variational inequality that are required by algorithms must be satisfied by the gradient map $x \mapsto \nabla_y f(x, x)$.

The next equivalent Ky Fan inequality plays a key role in many solution methods.

Corollary 2.2.10 *Suppose $f(x, \cdot)$ is τ-convex for any $x \in C$ with $\tau \geq 0$ and let*

$$f_\alpha(x, y) = f(x, y) + \alpha \|x - y\|^2 / 2 \qquad (2.16)$$

with $\alpha \geq -\tau$. Then, EP(f, C) and EP(f_α, C) have the same set of solutions.

Proof Since $y \mapsto \alpha \|x - y\|^2 / 2$ is α-convex, the function $f_\alpha(x, \cdot)$ is $(\tau + \alpha)$-convex by Corollary 2.1.7. Moreover, its subdifferential, see (2.4), is given by

$$\partial_y f_\alpha(x, y) = \partial_y f(x, y) + \alpha(y - x)$$

and in particular $\partial_y f_\alpha(x, x) = \partial_y f(x, x)$ holds for any $x \in C$ so that (2.14) holds.   □

Problem EP(f_α, C) is called *auxiliary Ky Fan inequality* and it is intensively exploited in Chap. 3. The equivalence between EP(f, C) and EP(f_α, C) allows deducing some alternative formulations of Theorems 2.2.1 and 2.2.2 when $\alpha > -\tau$. First, consider the multivalued map $Y_\alpha : \mathbb{R}^n \rightrightarrows \mathbb{R}^n$ given by

$$Y_\alpha(x) = \arg\min\{f_\alpha(x, y) : y \in C\}.$$

Since $f_\alpha(x, \cdot)$ is $(\tau + \alpha)$-convex with $\tau + \alpha > 0$, Theorem 2.1.9 guarantees that $Y_\alpha(x) = \{y_\alpha(x)\}$ is a singleton for any $x \in \mathbb{R}^n$ and the regularized gap function

$$\varphi_\alpha(x) = \max\{-f_\alpha(x, y) : y \in C\} = -\min\{f_\alpha(x, y) : y \in C\} \qquad (2.17)$$

satisfies $\varphi_\alpha(x) = -f_\alpha(x, y_\alpha(x))$. Pairing Corollary 2.2.10 with Theorems 2.2.1 and 2.2.2 leads to the following characterization.

Corollary 2.2.11 *Suppose $f(x, \cdot)$ is τ-convex for any $x \in C$ with $\tau \geq 0$. Given any $\alpha > -\tau$, the following statements are equivalent:*

a) *$\bar{x} \in \mathbb{R}^n$ solves EP(f, C),*
b) *$\bar{x} = y_\alpha(\bar{x})$,*
c) *$\bar{x} \in C$ and $\varphi_\alpha(\bar{x}) = 0$.*

Moreover, Theorems 2.2.6 and 2.2.7 can be exploited to prove that φ_α is continuously differentiable when C is a compact set.

Corollary 2.2.12 *Suppose C is compact, f is continuous and $f(\cdot, y)$ is differentiable for any $y \in \mathbb{R}^n$ with $\nabla_x f$ being continuous (in both variables). If $f(x, \cdot)$ is τ-convex for any $x \in C$ with $\tau \geq 0$, then the function $y_\alpha(\cdot)$ is continuous and φ_α is continuously differentiable with*

$$\nabla \varphi_\alpha(x) = -\nabla_x f(x, y_\alpha(x)) - \alpha(x - y_\alpha(x)) \qquad (2.18)$$

for any given $\alpha > -\tau$.

Another important merit function is the so-called *D-gap* function

$$\varphi_{\alpha\beta}(x) = \varphi_\alpha(x) - \varphi_\beta(x) \qquad (2.19)$$

where φ_α and φ_β are gap functions defined through (2.17) and the prefix D stands for difference. It allows reformulating the Ky Fan inequality as a unconstrained optimization problem.

Theorem 2.2.13 *Suppose C is compact, f is continuous and $f(\cdot, y)$ differentiable for any $y \in \mathbb{R}^n$ with $\nabla_x f$ being continuous (in both variables). If $f(x, \cdot)$ is τ-convex for any $x \in C$ with $\tau \geq 0$, then the following statements hold for any $\beta > \alpha > -\tau$:*

a) *$\varphi_{\alpha\beta}(x) \geq (\beta - \alpha)\|x - y_\beta(x)\|^2/2$ for any $x \in \mathbb{R}^n$,*
b) *$\varphi_{\alpha\beta}$ is a merit function on $\mathbb{R}^n$ for $EP(f, C)$,*
c) *$\varphi_{\alpha\beta}$ is continuously differentiable on $\mathbb{R}^n$ with*

$$\nabla \varphi_{\alpha\beta}(x) = \nabla_x f(x, y_\beta(x)) - \nabla_x f(x, y_\alpha(x)) - s(x) \qquad (2.20)$$

where $s(x) = \alpha(x - y_\alpha(x)) - \beta(x - y_\beta(x))$,

d) $\bar{x} \in \mathbb{R}^n$ solves $EP(f, C)$ if and only if $\bar{x} \in C$, $r(\bar{x}) = y_\alpha(\bar{x}) - y_\beta(\bar{x}) = 0$ and $s(\bar{x}) = 0$.

Proof Given any $x \in \mathbb{R}^n$ the unique minimizers $y_\alpha(x)$ and $y_\beta(x)$ of $f_\alpha(x, \cdot)$ and $f_\beta(x, \cdot)$ over C satisfy

$$
\begin{aligned}
\varphi_{\alpha\beta}(x) &= -f(x, y_\alpha(x)) - \alpha\|x - y_\alpha(x)\|^2/2 + f(x, y_\beta(x)) + \beta\|x - y_\beta(x)\|^2/2 \\
&\geq -f(x, y_\beta(x)) - \alpha\|x - y_\beta(x)\|^2/2 + f(x, y_\beta(x)) + \beta\|x - y_\beta(x)\|^2/2 \\
&= \frac{\beta - \alpha}{2}\|x - y_\beta(x)\|^2 \geq 0.
\end{aligned}
$$

Moreover, $\varphi_{\alpha\beta}(\bar{x}) = 0$ if and only if $\bar{x} = y_\beta(\bar{x})$, that is $\bar{x}$ solves EP(f, C) by Corollary 2.2.11. Hence, $\varphi_{\alpha\beta}$ is a merit function on $\mathbb{R}^n$ for EP(f, C).

The differentiability of $\varphi_{\alpha\beta}$ readily follows from the differentiability of φ_α and φ_β, which is guaranteed by Corollary 2.2.12, while (2.20) follows from (2.18).

Finally, given any $\bar{x} \in \mathbb{R}^n$, $r(\bar{x}) = s(\bar{x}) = 0$ holds if and only if $y_\alpha(\bar{x}) = y_\beta(\bar{x}) = \bar{x} \in C$, i.e., $\bar{x}$ solves EP(f, C) by Corollary 2.2.11.   □

Beyond the variational inequality (2.15) and the auxiliary problem EP(f_α, C), the Ky Fan inequality EP(f, C) is strongly related to another inequality known as the *Minty inequality* that reads

$$\text{find } \bar{y} \in C \text{ such that } f(x, \bar{y}) \leq 0 \text{ for all } x \in C. \qquad \text{MEP}(f, C)$$

Clearly, the pseudomonotonicity of f is enough to guarantee that any solution of EP(f, C) solves MEP(f, C) as well. The vice versa holds for local solutions of MEP(f, C).

Definition 2.2.14 A point $\bar{y} \in C$ is a *local solution* of MEP(f, C) if there exists $r > 0$ such that

$$f(x, \bar{y}) \leq 0 \text{ for all } x \in C \cap B(\bar{y}, r).$$

Theorem 2.2.15 *Suppose $f(x, \cdot)$ is convex for any $x \in C$ and $f(\cdot, y)$ is upper semicontinuous for any $y \in C$. Then, any local solution of MEP(f, C) solves EP(f, C). If, in addition, f is pseudomonotone, then the solution sets of EP(f, C) and MEP(f, C) coincide and they are closed and convex.*

Proof Let $\bar{y} \in C$ be a local solution of MEP(f, C) and $r > 0$ the corresponding radius. Consider any $y \in C$ with $y \neq \bar{y}$ and $\hat{y} = (1 - s)\bar{y} + sy$ with $s > 0$ small enough to have $\hat{y} \in C \cap B(\bar{y}, r)$. Setting $z_t = (1 - t)\bar{y} + t\hat{y}$, then $f(z_t, \bar{y}) \leq 0$ holds for any $t \in (0, 1)$. Since $f(z_t, \cdot)$ is convex, then

$$0 = f(z_t, z_t) \leq (1 - t)f(z_t, \bar{y}) + tf(z_t, \hat{y})$$

implies that $f(z_t, \hat{y}) \geq 0$ holds for any $t \in (0, 1)$ and

$$f(\bar{y}, \hat{y}) \geq \limsup_{t \to 0^+} f(z_t, \hat{y}) \geq 0$$

follows since $f(\cdot, \hat{y})$ is upper semicontinuous. Moreover, $\bar{y}$ and y satisfy

$$0 \leq f(\bar{y}, \hat{y}) \leq (1 - s)f(\bar{y}, \bar{y}) + sf(\bar{y}, y) = sf(\bar{y}, y)$$

thanks to the convexity of $f(\bar{y}, \cdot)$. Hence, $\bar{y}$ solves EP(f, C). As pseudomonotonicity guarantees that any solution of EP(f, C)solves MEP(f, C) as well, the sets of the solutions of the two problems coincide.

Let $\{x^k\}$ be a sequence of solutions of EP(f, C) such that $x^k \to x$ for some $x \in C$. Therefore, the upper semicontinuity $f(\cdot, y)$ guarantees that

$$f(x, y) \geq \limsup_{k \to +\infty} f(x^k, y) \geq 0$$

holds for any $y \in C$ so that x solves EP(f, C) as well. Thus, the set of the solutions of EP(f, C) is closed. Its convexity can be checked through the equivalent Minty inequality. Indeed, given any two solutions y_1, y_2 of MEP(f, C), $t \in [0, 1]$ and $x \in C$, the convexity of $f(x, \cdot)$ implies

$$f(x, ty_1 + (1 - t)y_2) \leq tf(x, y_1) + (1 - t)f(x, y_2) \leq 0,$$

that is $ty_1 + (1 - t)y_2$ solves MEP(f, C).   □

Notice that local solutions of the Minty inequality are not necessarily solutions of the Minty inequality over the whole feasible region as the following example shows.

Example 2.2.16 The Minty inequality MEP(f, C) with $f(x, y) = (x-y)(x-y-1)$ and $C = [0, 2]$ does not have any solution. Indeed, $y \in (0, 2]$ implies $f(y/2, y) = y(y + 2)/4 > 0$ while $f(2, 0) = 2 > 0$. On the contrary, $\bar{y} = 0$ is a local solution since $f(x, 0) = x(x - 1) \leq 0$ holds for any $x \in [0, 1] = C \cap B(0, 1)$.

2.3 Existence

The set of solutions of a Ky Fan inequality can be viewed as the intersection of a (possibly uncountable) family of sets. More precisely, $\bar{x} \in C$ solves EP(f, C) if and only if

$$\bar{x} \in \bigcap_{y \in C} T(y), \tag{2.21}$$

where $T : \mathbb{R}^n \rightrightarrows \mathbb{R}^n$ is the multivalued map given by

$$T(y) = \{x \in C : f(x, y) \geq 0\}.$$

Therefore, geometrical tools such as the Knaster–Kuratowski–Mazurkiewicz lemma that couples finite inclusions with finite intersections of sets are useful to prove the existence of solutions.

Lemma 2.3.1 (KKM)   *Let I be a finite set of indices, $\{x^i\}_{i \in I} \subseteq \mathbb{R}^n$ and $\{F_i\}_{i \in I}$ be a family of closed subsets of $\mathbb{R}^n$. If the inclusion*

$$\mathrm{conv}\left(\{x^i\}_{i \in J}\right) \subseteq \bigcup_{i \in J} F_i \tag{2.22}$$

holds for any $J \subseteq I$, then the intersection

$$\mathrm{conv}\left(\{x^i\}_{i \in I}\right) \cap \bigcap_{i \in I} F_i \tag{2.23}$$

is nonempty and compact.

Proof   The intersection (2.23) is compact since $\mathrm{conv}\left(\{x^i\}_{i \in I}\right)$ is compact. As for its nonemptiness, it is enough to prove it under the additional assumption that the points $\{x^i\}_{i \in I}$ are affinely independent.

Indeed, suppose they are not and choose any $m + 1$ affinely independent points $\{z^i\}_{i \in I}$ in $\mathbb{R}^m$ where $m + 1 = |I|$. Given any $z \in \mathrm{conv}\left(\{z^i\}_{i \in I}\right)$, set $\rho(z) = \sum_{i \in I} \lambda_i x^i$ where λ_i are the barycentric coordinates of z. The function ρ maps $\mathrm{conv}\left(\{z^i\}_{i \in I}\right)$ into $\mathrm{conv}\left(\{x^i\}_{i \in I}\right)$ and it is continuous since the barycentric coordinates are such. Consider the sets $E_i = \rho^{-1}(\mathrm{conv}\left(\{x^i\}_{i \in I}\right) \cap F_i) \subseteq \mathrm{conv}\left(\{z^i\}_{i \in I}\right)$, which are closed as they are the inverse images of closed sets by a continuous function. Given any $J \subseteq I$, the inclusion

$$\mathrm{conv}\left(\{z^i\}_{i \in J}\right) \subseteq \bigcup_{i \in J} E_i$$

holds since both $\mathrm{conv}\left(\{z^i\}_{i \in J}\right) \subseteq \rho^{-1}(\mathrm{conv}\left(\{x^i\}_{i \in J}\right))$ and (2.22) hold. If the thesis is true for sets of affinely independent points, then there exists some $z \in \bigcap_{i \in I} E_i$, that reads $\rho(z) \in \mathrm{conv}\left(\{x^i\}_{i \in I}\right) \cap F_i$ for any $i \in I$, i.e., $\rho(z)$ belongs to (2.23).

Thus, suppose $\{x^i\}_{i \in I}$ are affinely independent, so that their convex hull is an m-simplex. Consider a sequence of simplicial subdivisions $\{\mathscr{P}_k\}$ with $\mathrm{diam}(\mathscr{P}_k) \to 0$ as $k \to +\infty$. Given any $x \in V(\mathscr{P}_k)$, i.e., any vertex of the cells of $\mathscr{P}$, and its barycentric coordinates $\{\lambda_i\}_{i \in I}$, let $J = \{i \in I : \lambda_i > 0\}$ denote the smallest subset of I such that $x \in \mathrm{conv}\left(\{x^i\}_{i \in J}\right)$. This convex hull is a subset of $\bigcup_{i \in J} F_i$ and hence $x \in F_\ell$ for some $\ell \in J$. Setting $v_k(x) = \ell$, the function $v_k : V(\mathscr{P}_k) \to I$ turns out to be a Sperner proper labeling. Hence, there exists a completely labeled cell

$C_k \in \mathscr{P}_k$ by Theorem A.1.1. Let $\{\hat{x}^{i,k}\}_{i\in I}$ be the set of the vertices of C_k, which are listed according to the complete labeling, i.e., $v_k(\hat{x}^{i,k}) = i$ holds. All the sequences $\{\hat{x}^{i,k}\}_{k\in\mathbb{N}}$ are bounded since each belongs to the compact set conv $(\{x^i\}_{i\in I}) \cap F_i$. Therefore, possibly taking a suitable subsequence, any $i \in I$ satisfies $\hat{x}^{i,k} \to \hat{x}^i$ as $k \to +\infty$ for some $\hat{x}^i \in$ conv $(\{x^i\}_{i\in I}) \cap F_i$. Moreover, the inequalities

$$\|\hat{x}^i - \hat{x}^j\| \le \|\hat{x}^i - x^{i,k}\| + \|x^{i,k} - x^{j,k}\| + \|x^{j,k} - \hat{x}^j\|$$
$$\le \|\hat{x}^i - x^{i,k}\| + \mathrm{diam}(\mathscr{P}_k) + \|x^{j,k} - \hat{x}^j\|$$

guarantee $\hat{x}^i = \hat{x}^j$ for all $i, j \in I$ since all the terms in the second inequality go to zero as $k \to +\infty$. Therefore, the common cluster point belongs to the intersection of all the sets conv $(\{x^i\}_{i\in I}) \cap F_i$, i.e., (2.23) is nonempty. $\qquad\square$

A multivalued map $T : \mathbb{R}^n \rightrightarrows \mathbb{R}^n$ is said to satisfy the *KKM property* on a given $C \subseteq \mathbb{R}^n$ if the inclusion

$$\mathrm{conv}\left(\{x^i\}_{i\in I}\right) \subseteq \bigcup_{i\in I} T(x^i)$$

holds for any finite collection $\{x^i\}_{i\in I} \subseteq C$. In order to prove that the intersection of the images of such a map is nonempty by exploiting the KKM Lemma 2.3.1, the compactness of at least one image is needed as well.

Lemma 2.3.2 (Fan-KKM) *Suppose the multivalued map $T : \mathbb{R}^n \rightrightarrows \mathbb{R}^n$ is closed-valued. If it satisfies the KKM property on $C \subseteq \mathbb{R}^n$ and there exists $\bar{x} \in C$ such that $T(\bar{x})$ is compact, then the intersection $\bigcap_{x\in C} T(x)$ is nonempty and compact.*

Proof The sets $F(x) = T(x) \cap T(\bar{x})$ are closed subsets of $T(\bar{x})$ for any $x \in C$. Moreover, the KKM property guarantees that the intersection $\bigcap_{i\in I} F(x^i)$ is nonempty for any finite index set I and any collection of points $\{x^i\}_{i\in I} \subseteq C$ by the KKM Lemma 2.3.1, i.e., the family $\{F(x)\}_{x\in C}$ has the finite intersection property. Since $T(\bar{x})$ is compact, the thesis follows (see Appendix A.1). $\qquad\square$

Notice that Fan-KKM Lemma 2.3.2 does not guarantee that the intersection of images $\bigcap_{x\in C} T(x)$ contains some point of C even when C is convex but not closed.

Example 2.3.3 Let $C = (0, 1] \subseteq \mathbb{R}$ and $T(x) = [0, x]$ for any $x \in C$. All $T(x)$ are compact (and convex) and any sequence $0 < x^1 \le x^2 \le \ldots \le x^m \le 1$ satisfies

$$\mathrm{conv}\left(\{x^i\}_{i\in\{1,\ldots,m\}}\right) = [x^1, x^m] \subseteq [0, x^m] = \bigcup_{i=1}^m T(x^i).$$

Therefore all the assumptions of the Fan-KKM Lemma 2.3.2 holds and

$$\bigcap_{x\in C} T(x) = \bigcap_{x\in C} [0, x] = \{0\}$$

which does not belong to C.

Indeed, the closedness of C guarantees that $T \cap C : x \rightrightarrows T(x) \cap C$ satisfies all the assumptions of Lemma 2.3.2 and hence

$$C \cap \bigcap_{x \in C} T(x) = \bigcap_{x \in C} [C \cap T(x)] \neq \emptyset.$$

The Fan-KKM Lemma 2.3.2 allows achieving the existence of a solution of the Ky Fan inequality EP(f, C) under suitable convexity and continuity assumptions on the bifunction f.

Theorem 2.3.4 *Suppose $f(x, \cdot)$ is quasiconvex on C for any $x \in C$ and $f(\cdot, y)$ is upper semicontinuous for any $y \in C$. If C is compact, then EP(f, C) has at least one solution.*

Proof It is enough to show that the intersection in (2.21) is nonempty.

The upper semicontinuity of $f(\cdot, y)$ guarantees that $T(y)$ is closed and therefore it is also compact since it is a subset of C. The KKM property holds for T on C thanks to the quasiconvexity of the functions $f(x, \cdot)$. Indeed, given any finite set of points $\{x^i\}_{i \in I} \subseteq C$, any $x \in \text{conv}\left(\{x^i\}_{i \in I}\right)$ satisfies

$$0 = f(x, x) \leq \max\{f(x, x^i) : i \in I\}$$

and any $j \in I$ with $f(x, x^j) \geq 0$ guarantees $x \in T(x^j)$.

Therefore, the Fan-KKM Lemma 2.3.2 guarantees the nonemptiness of the intersection in (2.21). □

Theorem 2.3.4 requires the compactness of C. Anyway, it can be replaced by the following coercivity condition

$$\exists r > 0 \text{ s.t. } \forall x \in C \text{ with } \|x\| > r, \ \exists y \in C \text{ with } \|y\| < \|x\| \text{ and } f(x, y) \leq 0 \tag{2.24}$$

which is clearly satisfied for any large enough r when C is compact.

Theorem 2.3.5 *Suppose $f(x, \cdot)$ is convex on C for any $x \in C$ and $f(\cdot, y)$ is upper semicontinuous for any $y \in C$. If the coercivity condition (2.24) holds, then EP(f, C) has at least one solution.*

Proof Suppose C is not compact so that $C_k = C \cap \bar{B}(0, r_k)$ is nonempty, convex and compact for any given sequence $r_k \uparrow +\infty$ with $r_k > r$. Theorem 2.3.4 guarantees the existence of $x^k \in C_k$ that solves EP(f, C_k).

If $\|x^k\| > r$ holds for some $k \in \mathbb{N}$, then the coercivity condition (2.24) provides $y^k \in C_k$ with $\|y^k\| < r_k$ such that $f(x^k, y^k) \leq 0$ and thus $f(x^k, y^k) = 0$ holds as well since x^k solves EP(f, C_k). The convexity of $f(x^k, \cdot)$ guarantees that x^k solves EP(f, C) as well. Indeed, the inequalities

$$f(x^k, y) \geq t^{-1}[f(x^k, ty + (1-t)y^k) - (1-t)f(x^k, y^k)] = t^{-1}f(x^k, ty + (1-t)y^k) \geq 0$$

hold for any $y \in C$ and any small enough $t \in (0, 1)$ since $ty + (1 - t)y^k \to y^k$ as $t \downarrow 0$, the convexity of C and $\|y^k\| < r_k$ guarantee $ty + (1 - t)y^k \in C_k$.

On the contrary, if $\{x^k\} \subseteq C \cap \bar{B}(0, r)$ holds, then the boundedness of the sequence guarantees $x^k \to \bar{x}$ for some $\bar{x} \in C \cap \bar{B}(0, r)$ provided that a suitable subsequence is taken. Any $y \in C$ satisfies $y \in C_k$ and hence $f(x^k, y) \geq 0$ whenever $r_k \geq \|y\|$. Therefore, the upper semicontinuity of $f(\cdot, y)$ guarantees

$$f(\bar{x}, y) \geq \limsup_{k \to +\infty} f(x^k, y) \geq 0,$$

so that $\bar{x}$ solves EP(f, C).   □

Replacing $f(x, y) \leq 0$ with the stronger requirement $f(x, y) < 0$ in (2.24) entails also the boundedness of the set of the solutions of the Ky Fan inequality.

Corollary 2.3.6 *Suppose $f(x, \cdot)$ is convex on C for any $x \in C$ and $f(\cdot, y)$ is upper semicontinuous for any $y \in C$. If the coercivity condition*

$$\exists r > 0 \text{ s.t. } \forall x \in C \text{ with } \|x\| > r, \ \exists y \in C \text{ with } \|y\| < \|x\| \text{ and } f(x, y) < 0 \tag{2.25}$$

holds, then the set of the solutions of EP(f, C) is nonempty and bounded.

Proof Since (2.25) implies (2.24), the nonemptiness of the set of solutions follows from Theorem 2.3.5. Moreover, (2.25) actually states that no $x \in C$ with $\|x\| > r$ may solve EP(f, C) so that the boundedness of the set of solutions readily follows.
                                                                                            □

The equilibrium bifunction f satisfies the coercivity condition (2.25) whenever strong pseudomonotonicity holds, which guarantees the uniqueness of the solution as well.

Theorem 2.3.7 *Suppose $f(x, \cdot)$ is convex for any $x \in C$. If f is μ-pseudomonotone on C with $\mu > 0$, then*

a) *the coercivity condition (2.25) holds;*
b) *EP(f, C) has at most one solution;*
c) *if in addition $f(\cdot, y)$ is upper semicontinuous for any $y \in C$, then EP(f, C) has exactly one solution.*

Proof

a) Suppose (2.25) does not hold. Let $y \in C$ and consider any $k \in \mathbb{N}$ with $k > \|y\|$. Then, there exists $x^k \in C$ with $\|x^k\| > k$ such that $f(x^k, y) \geq 0$. Moreover, $f(y, x^k) \leq -\mu\|x^k - y\|^2$ holds as well thanks to the μ-pseudomonotonicity of f. Any $y^* \in \partial_y f(y, y)$ satisfies

$$f(y, x^k) \geq f(y, y) + \langle y^*, x^k - y \rangle = \langle y^*, x^k - y \rangle.$$

Pairing the two inequalities, the Cauchy–Schwarz inequality leads to the contradiction

$$0 \geq \mu \|x^k - y\|^2 + \langle y^*, x^k - y \rangle \geq \mu \|x^k - y\|^2 - \|y^*\| \cdot \|x^k - y\| \to +\infty$$

as $k \to +\infty$, since $\mu > 0$.

b) Suppose both $x \in C$ and $y \in C$ solve EP(f, C). Then, $f(x, y) \geq 0$ holds and the μ-pseudomonotonicity of f guarantees

$$0 \leq f(y, x) \leq -\mu \|x - y\|^2$$

so that $x = y$ must hold since $\mu > 0$.

c) It follows pairing b) with Theorem 2.3.5.   □

The solution of the Ky Fan inequality EP(f, C) is unique also when weak monotonicity or weak concavity assumptions are satisfied by the equilibrium bifunction provided that strong convexity, i.e., $\tau > 0$, holds as well.

Theorem 2.3.8 Suppose $f(x, \cdot)$ is τ-convex for any $x \in C$ with $\tau \geq 0$ and $f(\cdot, y)$ is upper semicontinuous for any $y \in C$. If any of the following conditions holds:

a) f is μ-monotone on C with $\mu > -\tau$;
b) f is μ-pseudomonotone on C with $\mu > -\tau/2$;
c) $f(\cdot, y)$ is γ-concave for any $y \in C$ with $\gamma > -\tau$;

then EP(f, C) has exactly one solution.

Proof Consider any $x, y \in C$. If f is μ-monotone, then $f_{-\tau}$ given as in (2.16) satisfies

$$f_{-\tau}(x, y) + f_{-\tau}(y, x) = f(x, y) + f(y, x) - \tau \|y - x\|^2 \leq -(\mu + \tau)\|y - x\|^2,$$

i.e., $f_{-\tau}$ is $(\mu + \tau)$-monotone with $\mu + \tau > 0$. If f is μ-pseudomonotone, then $f_{-\tau}(x, y) \geq 0$ or equivalently $f(x, y) \geq \tau \|x - y\|^2/2 \geq 0$ guarantees

$$f_{-\tau}(y, x) = f(x, y) - \tau \|x - y\|^2/2 \leq -(\mu + \tau/2)\|x - y\|^2,$$

i.e., $f_{-\tau}$ is $(\mu + \tau/2)$-pseudomonotone with $\mu + \tau/2 > 0$.

Since Corollary 2.1.7 guarantees that $f_{-\tau}(x, \cdot)$ is convex, the thesis under assumptions a) and b) follows immediately from Theorem 2.3.7 and Corollary 2.2.10.

Suppose $f(\cdot, y)$ is γ-concave. Chosen any α with $-\tau < \alpha < \gamma$, the function $f_\alpha(\cdot, y)$ is $(\gamma - \alpha)$-concave by Theorem 2.1.6. Therefore, Theorem 2.2.5 guarantees that φ_α is $(\gamma - \alpha)$-convex and continuous. Theorem 2.1.9 guarantees the existence of a unique minimizer $\bar{x}$ of φ_α on C. Let $\bar{y} = y_\alpha(\bar{x})$, $t \in (0, 1)$, $x_t = t\bar{y} + (1 - t)\bar{x}$ and

$y_t = y_\alpha(x_t)$. Since $f_\alpha(x_t, \cdot)$ is $(\alpha + \tau)$-convex, Theorem 2.1.9 with $g = f_\alpha(x_t, \cdot)$ implies

$$(\alpha + \tau)\|y_t - \bar{y}\|^2/2 \leq f_\alpha(x_t, \bar{y}) - f_\alpha(x_t, y_t) = f_\alpha(x_t, \bar{y}) + \varphi_\alpha(x_t).$$

The continuity of $f_\alpha(\cdot, \bar{y})$ and φ_α implies

$$\lim_{t \to 0^+} f_\alpha(x_t, \bar{y}) + \varphi_\alpha(x_t) = f_\alpha(\bar{x}, \bar{y}) + \varphi_\alpha(\bar{x}) = 0,$$

hence $y_t \to \bar{y}$ as $t \to 0^+$. The following chain of equalities and inequalities

$$
\begin{aligned}
0 &\leq \varphi_\alpha(x_t) - \varphi_\alpha(\bar{x}) \\
&= -f_\alpha(x_t, y_t) - \varphi_\alpha(\bar{x}) \\
&\leq -f_\alpha(x_t, y_t) + f_\alpha(\bar{x}, y_t) \\
&\leq -tf_\alpha(\bar{y}, y_t) - (1 - t)f_\alpha(\bar{x}, y_t) + f_\alpha(\bar{x}, y_t) \\
&= t[f_\alpha(\bar{x}, y_t) - f_\alpha(\bar{y}, y_t)]
\end{aligned}
$$

holds since y_t minimizes $f(x_t, \cdot)$ over C and $f_\alpha(\cdot, y_t)$ is concave. Therefore, any $t \in (0, 1)$ satisfies

$$f_\alpha(\bar{x}, y_t) \geq f_\alpha(\bar{y}, y_t).$$

Taking the limit as $t \to 0^+$ provides

$$-\varphi_\alpha(\bar{x}) = f_\alpha(\bar{x}, \bar{y}) \geq f_\alpha(\bar{y}, \bar{y}) = 0.$$

Therefore, $\varphi_\alpha(\bar{x}) = 0$ holds, i.e., $\bar{x}$ is the unique solution of EP(f, C) by Corollary 2.2.11.   □

The last part of the section is devoted to existence results that do not require any kind of convexity on the equilibrium bifunction. Convexity can be somehow replaced by a suitable kind of antimonotonicity.

Definition 2.3.9 An equilibrium bifunction f is said to be *cyclically monotone* on C if

$$f(x^0, x^1) + f(x^1, x^2) + \cdots + f(x^m, x^0) \leq 0$$

holds for any $m \in \mathbb{N}$ and any set of points $x^0, x^1, \ldots, x^m \in C$. If $-f$ is cyclically monotone on C, then f is called *cyclically antimonotone on C*.

Notice that cyclically (anti)monotone bifunctions are also (anti)monotone but not vice versa. Moreover, the following characterization of cyclical antimonotonicity provides an unexpected connection with optimization problems (see also Sect. 1.8).

Theorem 2.3.10 f *is cyclically antimonotone on C if and only if there exists a function g such that the inequality*

$$f(x, y) \geq g(y) - g(x) \tag{2.26}$$

holds for any $x, y \in C$.

Proof The "if" part condition is obvious. In order to prove the "only if" part, consider a given $x^0 \in C$ and the function

$$g(x) = -\inf\{f(x, x^m) + f(x^m, x^{m-1}) + \cdots + f(x^1, x^0)\}, \tag{2.27}$$

where the infimum is taken over all finite sets of points $\{x^1, x^2, \ldots, x^m\} \subset C$ for any $m \in \mathbb{N}$. Since f is cyclically antimonotone on C, any set of points $\{x^1, x^2, \ldots, x^m\}$ satisfies

$$f(x, x^m) + f(x^m, x^{m-1}) + \cdots + f(x^1, x^0) \geq -f(x^0, x)$$

together with any $x \in C$. As a consequence, $g(x) \leq f(x^0, x)$ holds and thus g is finite at any $x \in C$. Since the inequality

$$f(x, y) + f(y, x^m) + f(x^m, x^{m-1}) + \cdots + f(x^1, x^0) \geq -g(x)$$

holds for any $x, y \in C$ and any set of points $\{x^1, x^2, \ldots, x^m\} \subset C$, taking the infimum over all the set of points provides

$$f(x, y) - g(y) \geq -g(x)$$

and thus (2.26) holds with this choice of g. □

As a straightforward consequence of (2.26) any minimum point of g over C solves EP(f, C) whenever f is cyclically antimonotone. Therefore, existence results for optimization problems lead to existence results for Ky Fan inequalities.

Notice that g is lower semicontinuous if $f(\cdot, y)$ is upper semicontinuous for any $y \in C$ since it is the pointwise supremum of lower semicontinuous functions (see Appendix A.2). Hence, the following existence result for Ky Fan inequalities is an immediate consequence of Weierstrass Theorem A.2.1.

Theorem 2.3.11 *Suppose f is cyclically antimonotone on C and $f(\cdot, y)$ is upper semicontinuous for any $y \in C$. If C is compact, then EP(f, C) has at least one solution.*

Similarly to the convex case (see Theorems 2.3.4 and 2.3.5) the boundedness of C can be replaced by the coercivity condition (2.24).

Theorem 2.3.12 *Suppose f is cyclically antimonotone and $f(\cdot, y)$ is upper semi-continuous for any $y \in C$. If the coercivity condition (2.24) holds, then $EP(f, C)$ has at least one solution.*

Proof Consider the lower semicontinuous function g given by (2.27). It is enough to show that it has a minimum point over C. Weierstrass Theorem A.2.1 guarantees the existence of a minimum point $\bar{x}$ of g over $C_r = C \cap \bar{B}(0, r)$.

Suppose that $\bar{x}$ does not minimize g over C, so that there exists $x \in C$ with $\|x\| > r$ such that $g(x) < g(\bar{x})$. Consider the set

$$K = \{y \in C : g(y) \leq g(x)\} \cap \bar{B}(0, \|x\|),$$

which is clearly nonempty and compact. Therefore, the existence of a point $y \in K$ of least norm over K is guaranteed by Weierstrass Theorem A.2.1. Since $g(y) < g(\bar{x})$ holds, the optimality of $\bar{x}$ implies $y \notin C_r$. Hence, (2.24) guarantees the existence of some $z \in C$ such that $\|z\| < \|y\| \leq \|x\|$ and $g(z) \leq g(y) \leq g(x)$, that contradicts the choice of y. As a consequence, the assumption on $\bar{x}$ cannot hold and it minimizes g over C, which guarantees that it solves $EP(f, C)$ as well. □

2.4 Stability

The analysis of the behaviour of the set of solutions as some parameters of the problem (slightly) change is a key issue, for instance whenever data and/or their measurements are affected by uncertainty or errors. It would be desirable for the solutions to be slightly affected by small perturbations of the data, so that they can be somehow considered stable. In order to analyse this issue for Ky Fan inequalities, consider two parameters $\varepsilon \in \mathbb{R}^p$ and $\theta \in \mathbb{R}^q$, the former perturbing the equilibrium bifunction and the latter the feasible region. More precisely, consider $f : \mathbb{R}^n \times \mathbb{R}^n \times \mathbb{R}^p \to \mathbb{R}$ such that $f(x, x, \varepsilon) = 0$ for each $x \in \mathbb{R}^n$ and the multivalued map $C : \mathbb{R}^q \rightrightarrows \mathbb{R}^n$ that together provide the perturbed Ky Fan inequality

$$\text{find } \bar{x} \in C(\theta) \text{ such that } f(\bar{x}, y, \varepsilon) \geq 0 \text{ for all } y \in C(\theta) \tag{2.28}$$

for any fixed $\varepsilon \in \mathbb{R}^p$ and $\theta \in \mathbb{R}^q$.

Let $S : \mathbb{R}^p \times \mathbb{R}^q \rightrightarrows \mathbb{R}^n$ denote the multivalued map that provides the set of the solution of (2.28) as a function of $\varepsilon \in \mathbb{R}^p$ and $\theta \in \mathbb{R}^q$. Good stability properties require that S is at least closed.

Theorem 2.4.1 *If f is upper semicontinuous and C is a lower semicontinuous and closed multivalued map, then S is a closed multivalued map.*

Proof Consider a sequence $\{(\varepsilon^k, \theta^k, x^k)\} \subseteq$ gph S converging to some $(\bar{\varepsilon}, \bar{\theta}, \bar{x})$. The closedness of C implies $\bar{x} \in C(\bar{\theta})$. Given any $y \in C(\bar{\theta})$, the lower semicontinuity of C guarantees the existence of a subsequence $\{\theta^{k_\ell}\}$ of $\{\theta^k\}$ and a sequence $\{y^{k_\ell}\}$ with $y^{k_\ell} \in C(\theta^{k_\ell})$ converging to y (see Theorem A.3.6). Since $x^{k_\ell} \in S(\varepsilon^{k_\ell}, \theta^{k_\ell})$, then $f(x^{k_\ell}, y^{k_\ell}, \varepsilon^{k_\ell}) \geq 0$ holds and the upper semicontinuity of f implies

$$f(\bar{x}, y, \bar{\varepsilon}) \geq \limsup_{\ell \to +\infty} f(x^{k_\ell}, y^{k_\ell}, \varepsilon^{k_\ell}) \geq 0$$

so that $\bar{x} \in S(\bar{\varepsilon}, \bar{\theta})$.   □

Additional results can be easily achieved when the Ky Fan inequality (2.28) admits a unique solution $x(\varepsilon, \theta)$ for any $(\varepsilon, \theta) \in \mathbb{R}^p \times \mathbb{R}^q$. Monotonicity of the equilibrium bifunction and Lipschitz type behaviour of the data allow controlling how $x(\varepsilon, \theta)$ changes with ε and θ.

Theorem 2.4.2 *Given $\bar{\varepsilon} \in \mathbb{R}^p$ and $\bar{\theta} \in \mathbb{R}^q$, suppose there exist neighborhoods U of $\bar{\varepsilon}$ and V of $\bar{\theta}$ such that $V \subseteq$ dom C, $f(x, \cdot, \varepsilon)$ is convex and $f(\cdot, y, \varepsilon)$ is upper semicontinuous for any $x, y \in \mathbb{R}^n$ and any $\varepsilon \in U$. Moreover, suppose there exist $L_1, L_2, L_3, \mu > 0$ such that $f(\cdot, \cdot, \varepsilon)$ is μ-monotone for any $\varepsilon \in U$ and*

$$C(\theta) \subseteq C(\theta') + L_1 \|\theta - \theta'\| B(0, 1), \tag{2.29}$$

$$|f(x, y, \varepsilon) - f(x, y', \varepsilon)| \leq L_2 \|y - y'\|, \tag{2.30}$$

$$|f(x, y, \varepsilon) - f(x, y, \varepsilon')| \leq L_3 \|\varepsilon - \varepsilon'\| \cdot \|x - y\| \tag{2.31}$$

hold for any $x, y, y' \in \mathbb{R}^n$, $\varepsilon, \varepsilon' \in U$, $\theta, \theta' \in V$. Then, there exists a unique solution $x(\varepsilon, \theta)$ of (2.28) for any $(\varepsilon, \theta) \in U \times V$ and the following estimate

$$\|x(\varepsilon, \theta) - x(\varepsilon', \theta')\| \leq \frac{1}{\mu} \left(L_3 \|\varepsilon - \varepsilon'\| + \sqrt{2L_1 L_2 \mu} \|\theta - \theta'\|^{1/2} \right)$$

holds for any $(\varepsilon, \theta), (\varepsilon', \theta') \in U \times V$.

Proof Theorem 2.3.8 guarantees the existence of a unique solution $x(\varepsilon, \theta)$ of (2.28) for any $(\varepsilon, \theta) \in U \times V$.

Given any $\varepsilon, \varepsilon' \in U$ and any $\theta, \theta' \in V$, the triangle inequality provides

$$\|x(\varepsilon, \theta) - x(\varepsilon', \theta')\| \leq \|x(\varepsilon, \theta) - x(\varepsilon', \theta)\| + \|x(\varepsilon', \theta) - x(\varepsilon', \theta')\|. \tag{2.32}$$

The following chain of inequalities holds

$$
\begin{aligned}
0 &\le f(x(\varepsilon,\theta), x(\varepsilon',\theta), \varepsilon) + f(x(\varepsilon',\theta), x(\varepsilon,\theta), \varepsilon') \\
&= f(x(\varepsilon,\theta), x(\varepsilon',\theta), \varepsilon) + f(x(\varepsilon',\theta), x(\varepsilon,\theta), \varepsilon) \\
&\quad + f(x(\varepsilon',\theta), x(\varepsilon,\theta), \varepsilon') - f(x(\varepsilon',\theta), x(\varepsilon,\theta), \varepsilon) \\
&\le -\mu\|x(\varepsilon,\theta) - x(\varepsilon',\theta)\|^2 + L_3\|\varepsilon - \varepsilon'\| \cdot \|x(\varepsilon,\theta) - x(\varepsilon',\theta)\|,
\end{aligned}
$$

where the first is true since $x(\varepsilon,\theta)$ and $x(\varepsilon',\theta)$ solve (2.28) for the corresponding parameters, and the second follows from μ-monotonicity and (2.31). Therefore, the estimate

$$
\|x(\varepsilon,\theta) - x(\varepsilon',\theta)\| \le (L_3/\mu)\|\varepsilon - \varepsilon'\| \tag{2.33}
$$

holds. Inclusion (2.29) guarantees the existence of $y \in C(\theta)$ and $y' \in C(\theta')$ such that

$$
\|y - x(\varepsilon',\theta')\| \le L_1\|\theta - \theta'\| \quad \text{and} \quad \|y' - x(\varepsilon',\theta)\| \le L_1\|\theta - \theta'\|. \tag{2.34}
$$

The following chain of inequalities holds

$$
\begin{aligned}
\mu\|x(\varepsilon',\theta) - x(\varepsilon',\theta')\|^2 &\le -f(x(\varepsilon',\theta), x(\varepsilon',\theta'), \varepsilon') - f(x(\varepsilon',\theta'), x(\varepsilon',\theta), \varepsilon') \\
&\le f(x(\varepsilon',\theta'), y', \varepsilon') - f(x(\varepsilon',\theta'), x(\varepsilon',\theta), \varepsilon') \\
&\quad + f(x(\varepsilon',\theta), y, \varepsilon') - f(x(\varepsilon',\theta), x(\varepsilon',\theta'), \varepsilon') \\
&\le L_2(\|y' - x(\varepsilon',\theta)\| + \|y - x(\varepsilon',\theta')\|) \\
&\le 2L_1 L_2\|\theta - \theta'\|,
\end{aligned}
$$

where the first follows from μ-monotonicity, the second is true since $x(\varepsilon',\theta)$ and $x(\varepsilon',\theta')$ solve (2.28) for the corresponding parameters, the third follows from (2.30) and the last from (2.34). Therefore, the estimate

$$
\|x(\varepsilon',\theta) - x(\varepsilon',\theta')\| \le \sqrt{2L_1 L_2/\mu}\|\theta - \theta'\|^{1/2} \tag{2.35}
$$

holds. Therefore, (2.32), (2.33) and (2.35) provide the thesis. $\qquad\square$

The optimization problem

$$
\min\{g(x) : x \in C\}
$$

is said to be *Tikhonov well-posed* if it has a unique solution $\bar{x} \in C$ and every minimizing sequence $\{x^k\} \subseteq C$, i.e., such that $g(x^k) \to g(\bar{x})$, converges to $\bar{x}$. Informally speaking, points $x \in C$ with values $g(x)$ close to the minimum value are actually close to the unique optimal solution in well-posed optimization problems. Hence, well-posedness may play an important role in proving the convergence of minimization algorithms. A straightforward possibility to introduce well-posedness for Ky Fan inequalities relies on its equivalence with optimization problems through merit functions. Considering the gap function φ given by (2.10) leads to the following definition.

Definition 2.4.3 The Ky Fan inequality EP(f, C) is *Tikhonov well-posed* if it has a unique solution $\bar{x} \in C$ and any sequence $\{x^k\} \subseteq C$ such that $\varphi(x^k) \to 0$ converges to $\bar{x}$.

Pairing suitable convexity and continuity assumptions with strong pseudomonotonicity guarantees Tikhonov well-posedness.

Theorem 2.4.4 *Suppose* $f(x, \cdot)$ *is convex for any* $x \in C$, $f(\cdot, y)$ *is upper semicontinuous for any* $y \in C$ *and* f *is* μ-*pseudomonotone on* C *with* $\mu > 0$. *Then,* EP(f, C) *is Tikhonov well-posed.*

Proof Theorem 2.3.7 guarantees the uniqueness of the solution $\bar{x} \in C$. Let $\{x^k\} \subseteq C$ be a sequence such that $\varphi(x^k) \to 0$. The definition of φ and the μ-pseudomonotonicity of f guarantee the inequalities

$$\varphi(x^k) \geq -f(x^k, \bar{x}) \geq \mu\|x^k - \bar{x}\|^2 \geq 0.$$

Hence, $\varphi(x^k) \to 0$ implies $x^k \to \bar{x}$ as $k \to +\infty$. □

Approximate solutions allow characterising Tikhonov well-posedness. Let S_ε denote the set of all the ε-solutions of EP(f, C), that is $\bar{x}_\varepsilon \in S_\varepsilon$ if and only if $f(\bar{x}_\varepsilon, y) \geq -\varepsilon$ holds for any $y \in C$ or equivalently $\varphi(\bar{x}_\varepsilon) \leq \varepsilon$. Notice that the inclusion $S_{\varepsilon_1} \subseteq S_{\varepsilon_2}$ holds whenever $\varepsilon_1 < \varepsilon_2$ and $\bigcap_{\varepsilon > 0} S_\varepsilon$ coincides with the (possibly empty) set of the solutions of EP(f, C).

Theorem 2.4.5 *If* EP(f, C) *is Tikhonov well-posed, then* diam $S_\varepsilon \to 0$ *as* $\varepsilon \to 0$. *Vice versa,* EP(f, C) *is Tikhonov well-posed if* diam $S_\varepsilon \to 0$ *as* $\varepsilon \to 0$ *provided that* $S_\varepsilon \neq \emptyset$ *holds for any* $\varepsilon > 0$ *and* $f(\cdot, y)$ *is upper-semicontinuous for any* $y \in C$.

Proof By contradiction, suppose there exist $\alpha > 0$ and a sequence $\{\varepsilon_k\} \to 0$ such that diam $S_{\varepsilon_k} \geq 2\alpha$. Consider any $x^k, \hat{x}^k \in S_{\varepsilon_k}$ satisfying $\|x^k - \hat{x}^k\| \geq \alpha$. Since both $f(x^k, y) \geq -\varepsilon_k$ and $f(\hat{x}^k, y) \geq -\varepsilon_k$ hold for any $y \in C$, the values of the gap function φ at x^k and $\hat{x}^k$ satisfy $0 \leq \varphi(x^k) \leq \varepsilon_k$ and $0 \leq \varphi(\hat{x}^k) \leq \varepsilon_k$. Taking the limit as $k \to +\infty$ provides

$$\lim_{k \to +\infty} \varphi(x^k) = \lim_{k \to +\infty} \varphi(\hat{x}^k) = 0,$$

and thus $\{x^k\}$ and $\{\hat{x}^k\}$ are two minimizing sequences. The well-posedness of EP(f, C) guarantees $x^k \to \bar{x}$ and $\hat{x}^k \to \bar{x}$, where $\bar{x}$ is the unique solution of the Ky Fan inequality. Therefore, the contradiction

$$0 < \alpha \leq \|x^k - \hat{x}^k\| = \|x^k - \bar{x} + \bar{x} - \hat{x}^k\| \leq \|x^k - \bar{x}\| + \|\bar{x} - \hat{x}^k\| \to 0$$

readily follows.

Vice versa, let $\{x^k\} \subset C$ be a sequence such that $\varphi(x^k) \to 0$ as $k \to +\infty$. Given any $\alpha > 0$, there exists $\varepsilon > 0$ such that diam $S_\varepsilon < \alpha$ while $\varphi(x^k) \leq \varepsilon$ or equivalently $x^k \in S_\varepsilon$ hold for any $k > N$ for some suitable $N \in \mathbb{N}$. Hence, $\|x^k - x^h\| \leq \alpha$ holds for any pair $h, k > N$ so that $\{x^k\}$ is a Cauchy sequence. Thus, $x^k \to \bar{x}$ for some $\bar{x} \in C$. Moreover, the upper semicontinuity of $f(\cdot, y)$ guarantees that

$$f(\bar{x}, y) \geq \limsup_{k \to +\infty} f(x^k, y) \geq \limsup_{k \to +\infty} -\varphi(x_k) = 0$$

holds for any $y \in C$. Then, $\bar{x}$ is a solution of EP(f, C). Moreover, it is the unique solution since $\bar{x} \in \bigcap_{\varepsilon > 0} S_\varepsilon$ and diam $S_\varepsilon \to 0$. □

2.5 Error Bounds

This section provides some estimates of the distance of a (feasible or unfeasible) point from the set of the solutions of EP(f, C). Such error bounds rely on the value of the regularized gap function φ_α or the value of the D-gap function $\varphi_{\alpha\beta}$ at the considered point.

Throughout all the section $f(x, \cdot)$ is supposed to be τ-convex for any $x \in \mathbb{R}^n$ for some $\tau \geq 0$ that does not depend on the considered point. The following error bound for τ-convex functions is a key tool to get error bound for equilibria.

Lemma 2.5.1 *Suppose* $g : \mathbb{R}^n \to \mathbb{R}$ *is* τ-convex with $\tau \geq 0$. *Given* $\xi \in \mathbb{R}^n$ *and* $\alpha > -\tau$, *then the unique minimizer*

$$y^\star = \arg\min\{g(y) + \alpha\|y - \xi\|^2/2 : y \in C\}$$

and any $x \in C$ *satisfy*

$$g(x) + \alpha\|x - \xi\|^2/2 \geq g(y^\star) + \alpha\|y^\star - \xi\|^2/2 + (\alpha + \tau)\|x - y^\star\|^2/2. \quad (2.36)$$

Proof The objective function $g_\alpha(y) = g(y) + \alpha\|y - \xi\|^2/2$ is $(\tau + \alpha)$-convex (Theorem 2.1.6) with $\tau + \alpha > 0$ so that $y^\star$ is well defined by Theorem 2.1.9 and (2.36) readily follows from (2.8). □

The first result relies on monotonicity assumptions on the bifunction f.

Theorem 2.5.2 *Suppose $f(\cdot, y)$ is upper semicontinuous on C for any $y \in C$ and f is μ-monotone on C with $\mu > -\tau$. Given any $\alpha > -\tau$, there exists $M > 0$ such that*

$$\varphi_\alpha(x) \geq M \|x - \bar{x}\|^2 \tag{2.37}$$

holds for any $x \in C$, where $\bar{x}$ is the unique solution of EP(f, C).

Proof Theorem 2.3.8 guarantees the existence of a unique solution $\bar{x}$ of EP(f, C). Let $x \in C$ with $x \neq \bar{x}$. Any $s \in [0, 1]$ satisfies

$$\begin{aligned}
\varphi_\alpha(x) &\geq -f(x, s\bar{x} + (1-s)x) - \alpha \|s\bar{x} + (1-s)x - x\|^2/2 \\
&= -f(x, s\bar{x} + (1-s)x) - \alpha s^2 \|x - \bar{x}\|^2/2 \\
&\geq -sf(x, \bar{x}) - (1-s)f(x, x) + \tau s(1-s)\|x - \bar{x}\|^2/2 - \alpha s^2 \|x - \bar{x}\|^2/2 \\
&= sf(\bar{x}, x) - sf(\bar{x}, x) - sf(x, \bar{x}) + \left[\tau s(1-s) - \alpha s^2\right]\|x - \bar{x}\|^2/2 \\
&\geq sf(\bar{x}, x) + \left[s\mu + \frac{\tau}{2}s(1-s) - \frac{\alpha}{2}s^2\right]\|x - \bar{x}\|^2,
\end{aligned}$$

where the first inequality follows from the definition of φ_α, the second one from the τ-convexity of $f(x, \cdot)$, the third one from the μ-monotonicity of f.

Lemma 2.5.1 with $g = f(\bar{x}, \cdot)$, $\xi = \bar{x}$ and $y^\star = \bar{x}$ provides the inequality

$$f(\bar{x}, x) + \alpha \|x - \bar{x}\|^2/2 \geq f(\bar{x}, \bar{x}) + (\alpha + \tau)\|x - \bar{x}\|^2/2,$$

hence

$$f(\bar{x}, x) \geq \tau \|x - \bar{x}\|^2/2.$$

Therefore, the above chain of inequalities provides also

$$\varphi_\alpha(x) \geq \left[s\mu + \frac{\tau}{2}s(2-s) - \frac{\alpha}{2}s^2\right]\|x - \bar{x}\|^2$$

for any $s \in [0, 1]$. Taking the maximum over s, that is

$$M = \max_{s \in [0, 1]} \left[s\mu + \frac{\tau}{2}s(2-s) - \frac{\alpha}{2}s^2\right] = \begin{cases} \mu + (\tau - \alpha)/2 & \text{if } \alpha \in (-\tau, \mu), \\ \dfrac{(\mu + \tau)^2}{2(\alpha + \tau)} & \text{if } \alpha \geq \mu, \end{cases}$$

inequality (2.37) readily follows.   □

Monotonicity can be replaced by the weaker pseudomonotonicity provided the constant μ is not too small.

Theorem 2.5.3 *Suppose $f(\cdot, y)$ is upper semicontinuous on C for any $y \in C$ and f is μ-pseudomonotone on C with $\mu > -\tau/2$. Given any $\alpha \in (-\tau, 2\mu)$, the inequality*

$$\varphi_\alpha(x) \geq (\mu - \alpha/2)\|x - \bar{x}\|^2 \tag{2.38}$$

holds for any $x \in C$, where $\bar{x}$ is the unique solution of EP(f, C).

Proof Theorem 2.3.8 guarantees the existence of a unique solution $\bar{x}$ of EP(f, C). Given any $x \in C$, the inequalities

$$\varphi_\alpha(x) \geq -f(x, \bar{x}) - \alpha\|x - \bar{x}\|^2/2 \geq (\mu - \alpha/2)\|x - \bar{x}\|^2$$

follow from the definition of φ_α and the μ-pseudomonotonicity of f.   □

The gap function φ_α provides a further error bound on C when μ-(pseudo)monotonicity is replaced by a suitable concavity condition on $f(\cdot, y)$.

Theorem 2.5.4 *Suppose there exists $\gamma > -\tau$ such that $f(\cdot, y)$ is γ-concave for any $y \in C$. Given any $\alpha \in (-\tau, \gamma)$, the inequality*

$$\varphi_\alpha(x) \geq (\gamma - \alpha)\|x - \bar{x}\|^2/2$$

holds for any $x \in C$, where $\bar{x}$ is the unique solution of EP(f, C).

Proof Theorem 2.3.8 guarantees the existence of a unique solution $\bar{x}$ of EP(f, C). The function φ_α is $(\gamma - \alpha)$-convex by Theorem 2.2.5. Given any $x \in C$, the inequality

$$\varphi_\alpha(x) - (\gamma - \alpha)\|x\|^2/2 \geq \varphi_\alpha(\bar{x}) - (\gamma - \alpha)\|\bar{x}\|^2/2 + \langle x^* - (\gamma - \alpha)\bar{x}, x - \bar{x}\rangle$$

and equivalently

$$\varphi_\alpha(x) \geq \varphi_\alpha(\bar{x}) + \langle x^*, x - \bar{x}\rangle + (\gamma - \alpha)\|x - \bar{x}\|^2/2$$

hold for any $x^* \in \partial\varphi_\alpha(\bar{x})$ (see Sect. 2.1). Theorems 2.2.11 and 2.1.3 guarantee that $\varphi_\alpha(\bar{x}) = 0$ and $\langle \bar{x}^*, x - \bar{x}\rangle \geq 0$ holds for some $\bar{x}^* \in \partial\varphi_\alpha(\bar{x})$. Therefore, the conclusion follows.   □

Also the D-gap function $\varphi_{\alpha\beta}$ allows estimating the distance of any given point from the set of the solutions of EP(f, C).

Theorem 2.5.5 *Suppose f is continuously differentiable, $\nabla_x f(x, \cdot)$ is μ-monotone with $\mu > -\tau$, and $\nabla_y f$ is Lipschitz continuous with constant L. Given any $\beta > \alpha > -\tau$, there exists $M > 0$ such that*

$$\varphi_{\alpha\beta}(x) \geq M\|x - \bar{x}\|^2 \tag{2.39}$$

holds for any $x \in \mathbb{R}^n$, where $\bar{x}$ is the unique solution of EP(f, C).

Proof Given any $x, y \in C$, consider the function

$$g(t) := f(x + t(y - x), x) - f(x + t(y - x), y)$$

for $t \in [0, 1]$. Since f is continuously differentiable, g is derivable and

$$g'(t) = \langle \nabla_x f(x + t(y - x), x) - \nabla_x f(x + t(y - x), y), y - x \rangle \leq -\mu \|x - y\|^2$$

holds for any $t \in [0, 1]$, where the inequality follows from the μ-monotonicity of the map $\nabla_x f(x + t(y - x), \cdot)$. Therefore, the fundamental theorem of calculus provides

$$f(x, y) + f(y, x) = g(1) - g(0) = \int_0^1 g'(t) dt \leq -\mu \|y - x\|^2,$$

and thus f is μ-monotone on C. Theorem 2.3.8 guarantees the existence of a unique solution $\bar{x}$ of EP(f, C).

Since $f(z, \cdot)$ is τ-convex for any $z \in C$, the inequalities

$$f(x, \bar{x}) \geq f(x, x) + \langle \nabla_y f(x, x), \bar{x} - x \rangle + \tau \|x - \bar{x}\|^2 / 2,$$

$$f(\bar{x}, x) \geq f(\bar{x}, \bar{x}) + \langle \nabla_y f(\bar{x}, \bar{x}), x - \bar{x} \rangle + \tau \|x - \bar{x}\|^2 / 2$$

hold. Summing them provides

$$\langle \nabla_y f(\bar{x}, \bar{x}) - \nabla_y f(x, x), \bar{x} - x \rangle \geq -f(x, \bar{x}) - f(\bar{x}, x) + \tau \|x - \bar{x}\|^2$$
$$\geq (\mu + \tau) \|x - \bar{x}\|^2. \tag{2.40}$$

Let $x \in \mathbb{R}^n$ with $x \neq \bar{x}$. Since $\bar{x}$ solves EP(f, C), the equivalent variational inequality (see Sect. 2.2) gives

$$\langle \nabla_y f(\bar{x}, \bar{x}), y_\beta(x) - \bar{x} \rangle \geq 0.$$

Moreover, $y_\beta(x)$ minimizes $f_\beta(x, \cdot)$ over C so that Theorem 2.1.3 *c)* guarantees that

$$\langle \nabla_y f(x, y_\beta(x)) + \beta[y_\beta(x) - x], \bar{x} - y_\beta(x) \rangle \geq 0.$$

Summing the above two inequalities provides

$$0 \leq \langle \nabla_y f(\bar{x}, \bar{x}) - \nabla_y f(x, y_\beta(x)), y_\beta(x) - \bar{x} \rangle + \beta \langle y_\beta(x) - x, \bar{x} - y_\beta(x) \rangle$$
$$= \langle \nabla_y f(\bar{x}, \bar{x}) - \nabla_y f(x, x), y_\beta(x) - \bar{x} \rangle$$
$$+ \langle \nabla_y f(x, x) - \nabla_y f(x, y_\beta(x)), y_\beta(x) - \bar{x} \rangle + \beta \langle y_\beta(x) - x, \bar{x} - y_\beta(x) \rangle. \tag{2.41}$$

Therefore, the following chain of inequalities and equalities holds

$$(\mu + \tau)\|x - \bar{x}\|^2 \le \langle \nabla_y f(\bar{x}, \bar{x}) - \nabla_y f(x, x), \bar{x} - y_\beta(x) \rangle$$
$$+ \langle \nabla_y f(\bar{x}, \bar{x}) - \nabla_y f(x, x), y_\beta(x) - x \rangle$$
$$\le \langle \nabla_y f(x, x) - \nabla_y f(x, y_\beta(x)), y_\beta(x) - \bar{x} \rangle$$
$$+ \beta \langle y_\beta(x) - x, \bar{x} - y_\beta(x) \rangle + \langle \nabla_y f(\bar{x}, \bar{x}) - \nabla_y f(x, x), y_\beta(x) - x \rangle$$
$$= \langle \nabla_y f(x, x) - \nabla_y f(x, y_\beta(x)), y_\beta(x) - x \rangle$$
$$+ \langle \nabla_y f(x, x) - \nabla_y f(x, y_\beta(x)), x - \bar{x} \rangle$$
$$+ \langle \nabla_y f(\bar{x}, \bar{x}) - \nabla_y f(x, x), y_\beta(x) - x \rangle$$
$$+ \beta \langle y_\beta(x) - x, \bar{x} - x \rangle - \beta \|y_\beta(x) - x\|^2$$
$$\le \langle \nabla_y f(x, x) - \nabla_y f(x, y_\beta(x)), x - \bar{x} \rangle$$
$$+ \langle \nabla_y f(\bar{x}, \bar{x}) - \nabla_y f(x, x), y_\beta(x) - x \rangle$$
$$+ \beta \langle y_\beta(x) - x, \bar{x} - x \rangle - (\beta + \tau)\|x - y_\beta(x)\|^2$$
$$\le (L + \sqrt{2}L + \beta)\|x - y_\beta(x)\| \cdot \|x - \bar{x}\| - (\beta + \tau)\|x - y_\beta(x)\|^2$$
$$\le (L + \sqrt{2}L + \beta)\|x - y_\beta(x)\| \cdot \|x - \bar{x}\|$$

where the first inequality follows from (2.40), the second from inequality (2.41), the third from the monotonicity of the map $\nabla_y f(x, \cdot)$ (Corollary 2.1.12) and the fourth from the Lipschitz assumption. Setting $r = (\mu + \tau)/(L + \sqrt{2}L + \beta)$, the above chain and Theorem 2.2.13 imply

$$\varphi_{\alpha\beta}(x) \ge (\beta - \alpha)\|x - y_\beta(x)\|^2/2 \ge (\beta - \alpha)r^2\|x - \bar{x}\|^2/2,$$

i.e., (2.39) holds for $M = (\beta - \alpha)r^2/2$.   □

2.6 Notes and References

The chapter starts with a short collection of well-known results on convexity. The class of τ-convex functions and its properties have been widely studied in [127]. It is worth mentioning that the decomposition result of a τ-convex function into the sum of a convex function and a square (Theorem 2.1.6) is strictly related to the considered norm and it fails if $\|\cdot\|$ is not Hilbertian, i.e., it is not related to an inner product [47]. Various concepts of generalized monotonicity given for maps have been extended to bifunctions in [25, 76]. A systematic study of the relationships between several monotonicity conditions has been carried out in [31].

The gap function was originally introduced in [14] for variational inequalities and it yields an equivalent reformulation as a constrained optimization problem (see also [93]). Unfortunately, this function is rarely differentiable. In order to overcome this drawback, a regularized gap function has been proposed in [68] by adding a suitable strongly convex quadratic term. Subsequently, D-gap functions for variational inequalities have been introduced in [111, 129]. Gap and regularized gap functions for Ky Fan inequalities have been introduced in [92], while the reformulation as unconstrained optimization through D-gap functions has been studied in [83].

The continuity and/or the differentiability of gap functions is a basic requirement to devise iterative solution methods. The Maximum Theorem due to Berge [23] provides conditions for the upper semicontinuity of the set of solutions of a constrained parametric maximization problem and for the continuity of the value function. The differentiability of the value function has been described by Danskin in [55]. An interesting in-depth analysis on stability and sensitivity of perturbed optimization problems is given in the books [14, 40]. Theorems 2.2.6 and 2.2.7 arise as particular cases of these results. However, the common assumptions on f do not guarantee the uniqueness of minimizers and hence the practical utility of Theorem 2.2.7 is limited. Theorem 2.2.8 (see [44]) provides a sufficient condition for the equivalence of two equilibrium problems and it has interesting consequences. First of all, the reformulation of a differentiable equilibrium problem as the variational inequality (2.15) can be deduced from this result. Moreover, the equivalence between EP(f, C) and the auxiliary equilibrium problem EP(f_α, C) allows achieving a differentiable gap function (Corollary 2.2.12).

The Minty variational inequality has been introduced in [95] essentially as a tool for achieving existence results for variational inequalities. In fact, the equivalence between the two kinds of variational inequalities holds under the continuity and pseudomonotonicity of the operator. Local solutions of the Minty variational inequality have been originally considered in [15] where the authors provide an existence result for the variational inequality under a weak assumption of continuity and quasimonotonicity of the operator. These results have been extended to Ky Fan inequalities in [24] and refined in [45] (see Theorem 2.2.15).

The main existence result for EP(f, C) is due to Ky Fan [66] and it was actually stated in general topological vector spaces. This result has its roots in the Sperner lemma [123] about labeled simplicial subdivisions, which has been exploited by Knaster, Kuratowski and Mazurkiewicz to prove the so-called KKM lemma [75]. An equivalent version of KKM lemma (Lemma 2.3.2) was introduced by Ky Fan [65] and he exploited it to prove his existence result (Theorem 2.3.4) in [66]. Moreover, Theorem 2.3.4 implies the well-known Brouwer fixed-point theorem (see [41]), which in turn is known to imply Sperner lemma [130]. Therefore, all these results are equivalent (see Fig. 2.1). For a more detailed discussion on this issue see [41].

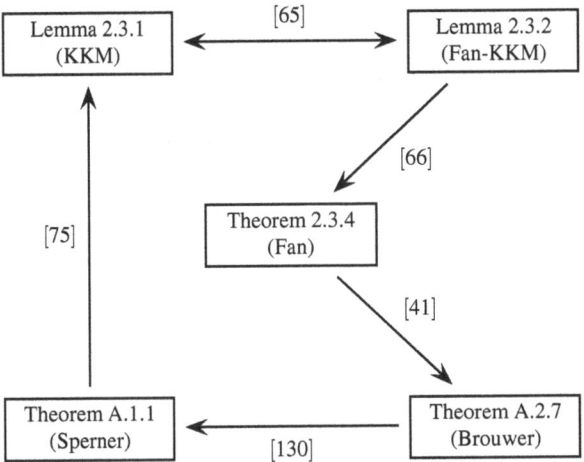

Fig. 2.1 A circular tour between KKM, Fan-KKM, Fan, Brouwer and Sperner theorems

Theorem 2.3.4 can be generalized along two different directions: either weakening the assumption of continuity of $f(\cdot, y)$ but adding some monotonicity condition (this approach is useful when the feasible set belongs to an infinite dimensional space) or replacing the boundedness of the feasible set with a suitable coercivity condition (see [43]). The coercivity conditions (2.24) and (2.25) were initially exploited in [24] for either pseudomonotone or quasimonotone bifunctions. A generalization of (2.24) has been proposed in [81] without any monotonicity assumption. However, suitable assumptions of monotonicity, which are often required in solution methods, guarantee the uniqueness of the solution as in Theorems 2.3.7 and 2.3.8 (see [33]). Currently, there are few existence results that do not ask for any kind of convexity on $f(x, \cdot)$ (see [26, 48]). Anyway, they usually require an additional strong assumption, called triangular property, which is rarely satisfied when the Ky Fan inequality is a variational inequality. Recently, a weaker property has been considered by introducing the class of cyclically antimonotone bifunctions [46], that allows obtaining the existence of both exact and approximate solutions.

As far as stability is concerned, the semicontinuity of the solution map of a Ky Fan inequality (Theorem 2.4.1) can be obtained exploiting semicontinuity results of the "infimal value function" [71]. Adding monotonicity assumptions that guarantee the uniqueness of the solution (Theorem 2.4.2), the solution map turns to be a Lipschitz function [1, 3, 5]. The Hölder continuity of the solution map may hold even if the solution is not unique [4]. Well-posedness can be introduced relying on merit functions and studied in the same fashion of optimization problems [27, 131]. The regular asymptotic behaviour of the solutions of a sequence of Ky Fan inequalities has been investigated in [39].

Merit functions play a key role also in the development of error bounds. Results based on gap functions have been considered in [33, 50, 82, 92]. Theorem 2.5.2 subsumes and extends the results of [82, 92] by adding the analysis of the strongly convex case ($\tau > 0$). A fairly new error bound involving pseudomonotonicity is given in Theorem 2.5.3. The approach of Theorem 2.5.4 based on concavity assumptions has been proposed in [33]. Error bounds using the D-gap function have been analysed in [52, 83, 135]. Theorem 2.5.5 has been inspired by [83] in the case $\tau = 0$. Actually, its proof shows clearly that the same result can be proved under the monotonicity of f rather than the monotonicity of gradient maps, which is a stronger assumption. Anyway, the latter plays a key role in the descent methods for D-gap functions (see Theorem 3.2.3).

Chapter 3
Algorithms for Equilibria

Algorithms for solving the Ky Fan inequality $EP(f, C)$ are tackled exploiting the tools developed in the previous chapter. Some assumptions on C and f hold throughout all the chapter in order to provide a unified algorithmic framework. Precisely, C is supposed to be convex and closed, f to be continuous and to satisfy $f(x, x) = 0$ for any $x \in \mathbb{R}^n$ while $f(x, \cdot)$ to be τ-convex for any $x \in \mathbb{R}^n$ for some $\tau \geq 0$ that does not depend upon the considered point x. Notice that this framework guarantees all the assumptions of the existence Theorems 2.3.4 and 2.3.8 except for the boundedness of C or some kind of monotonicity of f. Indeed, all the algorithms require at least one of them, so that the existence of a solution is always guaranteed.

3.1 Fixed-Point and Extragradient Methods

According to Theorem 2.2.1 the solutions of the Ky Fan inequality coincide with the fixed points of a multivalued map. Therefore, it seems reasonable to apply the classical algorithmic scheme for fixed points to this multivalued map in order to solve $EP(f, C)$. Actually, Corollary 2.2.11 provides a setting that is more useful from a computational point of view. In fact, the corollary shows the equivalence between the solutions of $EP(f, C)$ and the fixed points of a single-valued map, whose evaluation amounts to solving one strongly convex optimization problem. This basic version of the fixed point algorithm is given below.

Fixed point algorithm

Step 0.　Choose $x^0 \in C$, $\alpha > 0$ and set $k = 0$.
Step 1.　Compute $x^{k+1} = \arg\min\{f(x^k, y) + \alpha \|y - x^k\|^2/2 : y \in C\}$.
Step 2.　If $x^{k+1} = x^k$ then stop.
Step 3.　$k = k + 1$ and go to Step 1.

© Springer Nature Switzerland AG 2019
G. Bigi et al., *Nonlinear Programming Techniques for Equilibria*, EURO Advanced
Tutorials on Operational Research, https://doi.org/10.1007/978-3-030-00205-3_3

If the algorithm stops at Step 2, then Corollary 2.2.11 guarantees that x^k is a solution of EP(f, C). Otherwise, suitable assumptions on f guarantee that the infinite sequence of iterates $\{x^k\}$ converges to the unique solution $\bar{x}$ of EP(f, C) with linear rate of convergence, i.e., there exists a rate $M \in (0, 1)$ such that

$$\|x^{k+1} - \bar{x}\| \le M\|x^k - \bar{x}\|$$

holds for any $k \in \mathbb{N}$.

Theorem 3.1.1 *Suppose that f is μ-monotone and satisfies the inequality*

$$f(x, z) \le f(x, y) + f(y, z) + T_1\|y - x\|^2 + T_2\|y - z\|^2 \tag{3.1}$$

for any $x, y, z \in C$ for some $T_1, T_2 > 0$ with $\tau + \mu > T_1$. Chosen any $\alpha \ge 2T_2$, if the fixed point algorithm produces an infinite sequence $\{x^k\}$, the whole sequence converges to the unique solution of EP(f, C) with the linear rate of convergence

$$\sqrt{(\alpha + 2T_1 - 2\mu)/(\alpha + 2\tau)}. \tag{3.2}$$

Proof Theorem 2.3.8 guarantees that there exists a unique solution $\bar{x}$ of EP(f, C). Lemma 2.5.1 with $g = f(\bar{x}, \cdot)$ and $\xi = \bar{x}$ gives $y^* = \bar{x}$ and $x = x^{k+1}$ provides the inequality

$$f(\bar{x}, x^{k+1}) + \alpha\|x^{k+1} - \bar{x}\|^2/2 \ge f(\bar{x}, \bar{x}) + \alpha\|\bar{x} - \bar{x}\|^2/2 + (\alpha + \tau)\|x^{k+1} - \bar{x}\|^2/2,$$

hence

$$2f(\bar{x}, x^{k+1}) \ge \tau\|x^{k+1} - \bar{x}\|^2 \tag{3.3}$$

holds. Lemma 2.5.1 with $g = f(x^k, \cdot)$ and $\xi = x^k$ gives $y^* = x^{k+1}$ and $x = \bar{x}$ provides the inequality

$$f(x^k, \bar{x}) + \alpha\|\bar{x} - x^k\|^2/2 \ge f(x^k, x^{k+1}) + \alpha\|x^{k+1} - x^k\|^2/2 + (\alpha + \tau)\|x^{k+1} - \bar{x}\|^2/2.$$

Thus, the following chain of inequalities holds

$$(\alpha + \tau)\|x^{k+1} - \bar{x}\|^2 \le 2[f(x^k, \bar{x}) + f(\bar{x}, x^k) - f(\bar{x}, x^k) - f(x^k, x^{k+1})]$$
$$+ \alpha\|x^k - \bar{x}\|^2 - \alpha\|x^{k+1} - x^k\|^2$$
$$\le -2\mu\|x^k - \bar{x}\|^2 - 2f(\bar{x}, x^{k+1}) + 2T_1\|x^k - \bar{x}\|^2$$
$$+ 2T_2\|x^{k+1} - x^k\|^2 + \alpha\|x^k - \bar{x}\|^2 - \alpha\|x^{k+1} - x^k\|^2$$
$$\le (\alpha + 2T_1 - 2\mu)\|x^k - \bar{x}\|^2 + (2T_2 - \alpha)\|x^{k+1} - x^k\|^2$$

$$- \tau \|x^{k+1} - \bar{x}\|^2$$

$$\leq (\alpha + 2T_1 - 2\mu)\|x^k - \bar{x}\|^2 - \tau\|x^{k+1} - \bar{x}\|^2$$

where the second comes from the μ-monotonicity of f and inequality (3.1), the third from (3.3) and the last from the assumption on α. As a consequence, the inequality

$$\|x^{k+1} - \bar{x}\|^2 \leq \frac{\alpha + 2T_1 - 2\mu}{\alpha + 2\tau}\|x^k - \bar{x}\|^2$$

holds and the thesis follows since the above inequality guarantees that the constant is positive and $\tau + \mu > T_1$ guarantees that it is also smaller than 1. $\qquad\square$

Inequality (3.1) can be considered a Lipschitz-type condition. In fact, when EP(f, C) is a variational inequality, i.e., $f(x, y) = \langle F(x), y - x \rangle$ for some Lipschitz continuous map $F : \mathbb{R}^n \to \mathbb{R}^n$, (3.1) holds with $T_1 = T_2 = L/2$ where L is the Lipschitz constant. In fact, the Cauchy-Schwarz inequality guarantees

$$
\begin{aligned}
f(x, z) - f(x, y) - f(y, z) &= \langle F(x) - F(y), z - y \rangle \\
&\leq L\|y - x\| \cdot \|y - z\| \\
&\leq L\|y - x\|^2/2 + L\|y - z\|^2/2.
\end{aligned}
$$

Notice that Theorem 3.1.1 does not necessarily involve strong monotonicity, that is $\mu > 0$, provided that the constant τ of strong convexity is sufficiently large.

Exploiting Lemma 2.5.1 in a different way leads to a further convergence result, which requires pseudomonotonicity instead of monotonicity.

Theorem 3.1.2 *Suppose that f is μ-pseudomonotone and satisfies the inequality (3.1) for any $x, y, z \in C$ for some $T_1, T_2 > 0$ with $\tau + 2\mu > 2T_2$. Chosen any $\alpha \geq 2T_1$, if the fixed point algorithm produces an infinite sequence $\{x^k\}$, the whole sequence converges to the unique solution of EP(f, C) with the linear rate of convergence*

$$\sqrt{\alpha/(\alpha + \tau + 2\mu - 2T_2)}. \tag{3.4}$$

Proof Theorem 2.3.8 guarantees that there exists a unique solution $\bar{x}$ of EP(f, C). Applying Lemma 2.5.1 with $g = f(x^k, \cdot)$ and $\xi = x^k$ gives $y^\star = x^{k+1}$ and $x = \bar{x}$ provides

$$f(x^k, \bar{x}) + \alpha\|\bar{x} - x^k\|^2/2 \geq f(x^k, x^{k+1}) + \alpha\|x^{k+1} - x^k\|^2/2 + (\alpha + \tau)\|x^{k+1} - \bar{x}\|^2/2.$$

Thus

$$(\alpha + \tau)\|x^{k+1} - \bar{x}\|^2 \leq 2[f(x^k, \bar{x}) - f(x^k, x^{k+1})] + \alpha\|x^k - \bar{x}\|^2 - \alpha\|x^{k+1} - x^k\|^2$$

$$\leq 2f(x^{k+1}, \bar{x}) + 2T_1\|x^{k+1} - x^k\|^2 + 2T_2\|x^{k+1} - \bar{x}\|^2$$

$$+ \alpha \|x^k - \bar{x}\|^2 - \alpha \|x^{k+1} - x^k\|^2$$
$$\leq -2\mu \|x^{k+1} - \bar{x}\|^2 + 2T_1 \|x^{k+1} - x^k\|^2 + 2T_2 \|x^{k+1} - \bar{x}\|^2$$
$$+ \alpha \|x^k - \bar{x}\|^2 - \alpha \|x^{k+1} - x^k\|^2$$
$$\leq 2(T_2 - \mu) \|x^{k+1} - \bar{x}\|^2 + \alpha \|x^k - \bar{x}\|^2$$

where the second inequality comes from condition (3.1), the third from the μ-pseudomonotonicity of f and the last from the assumption on α. Therefore,

$$[\alpha + \tau + 2(\mu - T_2)] \|x^{k+1} - \bar{x}\|^2 \leq \alpha \|x^k - \bar{x}\|^2.$$

The assumption on α and the condition $\tau + 2\mu > 2T_2$ imply the thesis. $\square$

When EP(f, C) is a variational inequality with a Lipschitz operator F, then both $\tau = 0$ and $T_1 = T_2$ hold. Therefore, the rate (3.2) for the μ-monotone case is always smaller than the rate (3.4) for the μ-pseudomonotone case.

The assumptions of Theorems 3.1.1 and 3.1.2 imply the existence of a unique solution of EP(f, C). Fixed point techniques can be used under assumptions that do not guarantee the uniqueness of the solution as well. This can be done by solving two strongly convex optimization problems at each iteration: the resolution of the first problem acts as a prediction step, the second provides a correction step in which the objective function is updated while trying to remain close enough to the previous iterate.

Extragradient algorithm

Step 0. Choose $x^0 \in C$, $\alpha > 0$ and set $k = 0$.
Step 1. Compute $\bar{x}^k = \arg\min\{f(x^k, y) + \alpha \|y - x^k\|^2/2 \; : \; y \in C\}$.
Step 2. If $\bar{x}^k = x^k$ then stop.
Step 3. Compute $x^{k+1} = \arg\min\{f(\bar{x}^k, y) + \alpha \|y - x^k\|^2/2 \; : \; y \in C\}$.
Step 4. $k = k + 1$ and go to Step 1.

Notice that the stopping criterion at Step 2 is exactly the same of the fixed point algorithm.

Theorem 3.1.3 *Suppose there exists at least one solution $\hat{x}$ of MEP(f, C), f satisfies the inequality (3.1) for any $x, y, z \in C$ for some $T_1, T_2 > 0$. Chosen any $\alpha > \max\{2T_1, 2T_2 - \tau\}$, if the extragradient algorithm produces an infinite sequence $\{x^k\}$, then*

a) *the sequence has at least one cluster point and each of them is a solution of EP(f, C);*

b) *if f is also pseudomonotone, the whole sequence converges to one solution of EP(f, C);*

c) *if $\tau > 0$ holds, the whole sequence converges to one solution of $EP(f, C)$ with the linear rate of convergence*

$$\sqrt{\alpha/(\alpha + \tau)}. \tag{3.5}$$

Proof Applying Lemma 2.5.1 with $g = f(x^k, \cdot)$ and $\xi = x^k$ gives $y^\star = \bar{x}^k$ and $x = x^{k+1}$ provides

$$f(x^k, x^{k+1}) + \alpha \|x^{k+1} - x^k\|^2/2 \geq f(x^k, \bar{x}^k) + \alpha \|\bar{x}^k - x^k\|^2/2 + (\alpha + \tau)\|x^{k+1} - \bar{x}^k\|^2/2$$

while applying it with $g = f(\bar{x}^k, \cdot)$ and $\xi = x^k$ gives $y^\star = x^{k+1}$ and $x = \hat{x}$ provides

$$f(\bar{x}^k, \hat{x}) + \alpha \|\hat{x} - x^k\|^2/2 \geq f(\bar{x}^k, x^{k+1}) + \alpha \|x^{k+1} - x^k\|^2/2 + (\alpha + \tau)\|\hat{x} - x^{k+1}\|^2/2.$$

The sum of the two above inequalities provides the second inequality of the following chain

$$
\begin{aligned}
\alpha \|x^{k+1} - \hat{x}\|^2 &\leq (\alpha + \tau)\|x^{k+1} - \hat{x}\|^2 \\
&\leq 2[f(\bar{x}^k, \hat{x}) + f(x^k, x^{k+1}) - f(x^k, \bar{x}^k) - f(\bar{x}^k, x^{k+1})] \\
&\quad -\alpha \|\bar{x}^k - x^k\|^2 - (\alpha + \tau)\|x^{k+1} - \bar{x}^k\|^2 + \alpha \|x^k - \hat{x}\|^2 \\
&\leq (2T_1 - \alpha)\|\bar{x}^k - x^k\|^2 + (2T_2 - \alpha - \tau)\|x^{k+1} - \bar{x}^k\|^2 \\
&\quad + \alpha \|x^k - \hat{x}\|^2 \\
&\leq \alpha \|x^k - \hat{x}\|^2
\end{aligned}
\tag{3.6}
$$

where the third comes from the inequality (3.1) and the last from the assumptions on α. Therefore, the non-negative sequence $\{\|x^k - \hat{x}\|\}$ is monotone non-increasing and thus convergent. Its convergence guarantees also that the sequence $\{x^k\}$ is bounded and thus it has at least one cluster point. Let $\bar{x} \in C$ be one of them and consider the suitable subsequence $x^{k_\ell} \to \bar{x}$. Moreover, the inequalities (3.6) can be rearranged to get

$$0 \leq (\alpha - 2T_1)\|\bar{x}^k - x^k\|^2 + (\alpha + \tau - 2T_2)\|x^{k+1} - \bar{x}^k\|^2 \leq \alpha(\|x^k - \hat{x}\|^2 - \|x^{k+1} - \hat{x}\|^2)$$

so that $\|\bar{x}^k - x^k\| \to 0$ follows. Thus, $\bar{x}^{k_\ell} \to \bar{x}$ holds as well. Taking the limit in

$$\bar{x}^{k_\ell} = \arg\min\{f(x^{k_\ell}, y) + \alpha \|y - x^{k_\ell}\|^2/2 \ : \ y \in C\}$$

gives

$$\bar{x} = \arg\min\{f(\bar{x}, y) + \alpha \|y - \bar{x}\|^2/2 \ : \ y \in C\}.$$

Therefore, Corollary 2.2.11 guarantees that $\bar{x}$ is a solution of EP(f, C). If f is pseudomonotone, then $\bar{x}$ is a solution of MEP(f, C) as well and choosing $\hat{x} = \bar{x}$ the limit of the sequence $\{\|x^k - \hat{x}\|\}$ satisfies

$$\lim_{k \to +\infty} \|x^k - \bar{x}\| = \lim_{\ell \to +\infty} \|x^{k_\ell} - \bar{x}\| = 0,$$

i.e., the whole sequence $\{x^k\}$ converges to $\bar{x}$.

Finally, if $\tau > 0$, then the convergence of the sequence to one solution and its rate (3.5) follow immediately from (3.6), taking into account that $\hat{x}$ solves EP(f, C) (see Theorem 2.2.15). □

In order to get convergence when condition (3.1) is not satisfied, the correction step can be replaced by a suitable line search between x^k and $\bar{x}^k$ and a double projection. In order to simplify the presentation, f is supposed to be continuously differentiable for the remainder of the section.

Hyperplane extragradient algorithm

Step 0. Choose $x^0 \in C$, $\alpha > -\tau/2$, $\theta, \sigma \in (0, 1)$, let $\beta = \alpha + \tau/2$ and set $k = 0$.
Step 1. Compute $\bar{x}^k = \arg\min\{f(x^k, y) + \alpha\|y - x^k\|^2/2 \ : \ y \in C\}$.
Step 2. If $\bar{x}^k = x^k$ then stop.
Step 3. Find the smallest $m \in \mathbb{N}$ such that $\theta_k = \theta^m$ and $z^k = (1 - \theta_k)x^k + \theta_k \bar{x}^k$ satisfy

$$f(z^k, x^k) - f(z^k, \bar{x}^k) \geq \beta\sigma\|x^k - \bar{x}^k\|^2.$$

Step 4. Let $d^k = \nabla_y f(z^k, x^k)$, set $\delta_k = f(z^k, x^k)/\|d^k\|^2$ and compute

$$x^{k+1} = P_C(x^k - \delta_k d^k).$$

Step 5. $k = k + 1$ and go to Step 1.

The line search at Step 3 is well defined. Indeed, suppose

$$f((1 - \theta^m)x^k + \theta^m \bar{x}^k, x^k) - f((1 - \theta^m)x^k + \theta^m \bar{x}^k, \bar{x}^k) < \beta\sigma\|x^k - \bar{x}^k\|^2$$

holds for any $m \in \mathbb{N}$ so that taking the limit as $m \to +\infty$ provides the inequality

$$f(x^k, \bar{x}^k) + \beta\sigma\|x^k - \bar{x}^k\|^2 > 0.$$

On the contrary, Lemma 2.5.1 with $g = f(x^k, \cdot)$ and $\xi = x = x^k$ gives $y^\star = \bar{x}^k$ and

$$0 = f(x^k, x^k) + \alpha\|x^k - x^k\|^2/2 \geq f(x^k, \bar{x}^k) + \beta\|x^k - \bar{x}^k\|^2$$

so that a contradiction arises with the previous inequality since $x^k \neq \bar{x}^k$ and $\sigma < 1$.

Step 4 requires $d^k \neq 0$, that is $\nabla_y f(z^k, x^k) \neq 0$. Indeed, the opposite would guarantee x^k to minimize the convex function $f(z^k, \cdot)$ over $\mathbb{R}^n$ so that the inequality $0 = f(z^k, z^k) \geq f(z^k, x^k)$ would hold. On the contrary, the convexity of $f(z^k, \cdot)$ implies

$$0 = f(z^k, z^k) \leq (1 - \theta_k) f(z^k, x^k) + \theta_k f(z^k, \bar{x}^k)$$

that coupled with the inequality at Step 3 gives

$$f(z^k, x^k) \geq \theta_k [f(z^k, x^k) - f(z^k, \bar{x}^k)] \geq \theta_k \beta \sigma \|x^k - \bar{x}^k\|^2 > 0. \tag{3.7}$$

Step 3 amounts to finding an hyperplane

$$H^k = \{x \in \mathbb{R}^n \ : \ f(z^k, x^k) + \langle d^k, x - x^k \rangle = 0\}$$

that separates the current iterate from the solution set of EP(f, C). Indeed, the current iterate x^k satisfies

$$f(z^k, x^k) + \langle d^k, x^k - x^k \rangle = f(z^k, x^k) > 0$$

while any solution $\hat{x}$ of EP(f, C) satisfies

$$f(z^k, \hat{x}) \geq f(z^k, x^k) + \langle d^k, \hat{x} - x^k \rangle. \tag{3.8}$$

If f is pseudomonotone, then $f(z^k, \hat{x}) \leq 0$ holds and separation actually occurs.

The stepsize δ_k is chosen in such a way that $x^k - \delta_k d^k$ lies on the hyperlane H^k. In fact, the following equalities

$$f(z^k, x^k) + \langle d^k, x^k - \delta_k d^k - x^k \rangle = f(z^k, x^k) - \delta_k \|d^k\|^2 = 0$$

hold. Therefore, $x^k - \delta_k d^k$ is the projection of the current iterate x^k onto H^k and the new iterate x^{k+1} is the projection of such a projection onto the feasible region C.

Theorem 3.1.4 *Suppose C is compact and f is pseudomonotone. Chosen any $\alpha > -\tau/2$, if the hyperplane extragradient algorithm produces an infinite sequence $\{x^k\}$, then the whole sequence converges to one solution of EP(f, C).*

Proof Theorem 2.3.4 guarantees the existence of a solution $\hat{x}$ of EP(f, C). Thus, the pseudomonotonicity of f implies $f(z^k, \hat{x}) \leq 0$ and (3.8) yields

$$\langle d^k, x^k - \hat{x} \rangle \geq f(z^k, x^k) = \delta_k \|d^k\|^2.$$

Exploiting that the projection is non-expansive (see Theorem 2.1.4), the following chain of equalities and inequalities

$$
\begin{aligned}
\|x^{k+1} - \hat{x}\|^2 &= \|P_C(x^k - \delta_k d^k) - P_C(\hat{x})\|^2 \\
&\leq \|x^k - \delta_k d^k - \hat{x}\|^2 \\
&= \|x^k - \hat{x}\|^2 - 2\delta_k \langle d^k, x^k - \hat{x} \rangle + \delta_k^2 \|d^k\|^2 \\
&\leq \|x^k - \hat{x}\|^2 - \delta_k^2 \|d^k\|^2
\end{aligned} \tag{3.9}
$$

readily follows from the previous inequality. Therefore, the sequence $\{\|x^k - \hat{x}\|\}$ is monotone decreasing, bounded by below and thus convergent. Applying the above inequality iteratively provides

$$
\|x^{k+1} - \hat{x}\|^2 \leq \|x^0 - \hat{x}\|^2 - \sum_{\ell=0}^{k} \delta_\ell^2 \|d^\ell\|^2
$$

that allows concluding also

$$
\delta_k \|d^k\| = f(z^k, x^k)/\|d^k\| \to 0 \text{ as } k \to +\infty. \tag{3.10}
$$

The boundedness of $\{x^k\}$ guarantees the existence of at least one cluster point $\bar{x}$ of the sequence. If $\bar{x}$ is a solution of EP(f, C), then the whole sequence $\{x^k\}$ converges to $\bar{x}$ because of the convergence of $\{\|x^k - \hat{x}\|\}$ for $\hat{x} = \bar{x}$, and the thesis holds.

The boundedness of $\{x^k\}$ guarantees also that $\{\bar{x}^k\}$ is bounded since y_α is continuous (see Corollary 2.2.12) and hence $\{z^k\}$ is bounded as well. Therefore, there exist subsequences $\{x^{k_\ell}\}$, $\{\bar{x}^{k_\ell}\}$ and $\{z^{k_\ell}\}$ such that $x^{k_\ell} \to \bar{x}$, $\bar{x}^{k_\ell} \to \tilde{x}$ and $z^{k_\ell} \to \bar{z}$ for some suitable $\tilde{x}, \bar{z} \in C$. Taking the limit in

$$
\bar{x}^{k_\ell} = \arg\min\{f(x^{k_\ell}, y) + \alpha\|y - x^{k_\ell}\|^2/2 \; : \; y \in C\}
$$

gives

$$
\tilde{x} = \arg\min\{f(\bar{x}, y) + \alpha\|y - \bar{x}\|^2/2 \; : \; y \in C\}.
$$

Therefore, it is enough to prove $\tilde{x} = \bar{x}$ to get that $\bar{x}$ is a solution of EP(f, C) by Corollary 2.2.11. Two cases may occur: $\limsup \theta_{k_\ell} > 0$ or $\theta_{k_\ell} \to 0$ as $\ell \to +\infty$.

In the former case, $\theta_{k_\ell} \to \bar{\theta}$ for some $\bar{\theta} > 0$ (eventually taking a further subsequence). Since $d^{k_\ell} \to \nabla_y f(\bar{z}, \bar{x})$, then (3.10) implies $f(z^{k_\ell}, x^{k_\ell}) \to 0$ and (3.7) implies $\tilde{x} = \bar{x}$.

In the latter case, consider $\hat{z}^{k\ell} = (1 - \theta^{-1}\theta_{k\ell})x^{k\ell} + \theta^{-1}\theta_{k\ell}\bar{x}^{k\ell}$. By the choice of $\theta_{k\ell}$ the inequality

$$f(\hat{z}^{k\ell}, x^{k\ell}) - f(\hat{z}^{k\ell}, \bar{x}^{k\ell}) < \beta\sigma\|x^{k\ell} - \bar{x}^{k\ell}\|^2$$

holds for any k. Taking the limit gives

$$f(\bar{x}, \tilde{x}) + \beta\sigma\|\bar{x} - \tilde{x}\|^2 \geq 0$$

since $\hat{z}^{k\ell} \to \bar{x}$ and $f(\hat{z}^{k\ell}, x^{k\ell}) \to f(\bar{x}, \bar{x}) = 0$.

Lemma 2.5.1 with $g = f(\bar{x}, \cdot)$, $\xi = x = \bar{x}$ gives $y^\star = \tilde{x}$ and

$$0 = f(\bar{x}, \bar{x}) + \alpha\|\bar{x} - \bar{x}\|^2/2 \geq f(\bar{x}, \tilde{x}) + \beta\|\bar{x} - \tilde{x}\|^2.$$

Coupling this inequality with the previous provides

$$f(\bar{x}, \tilde{x}) + \beta\|\bar{x} - \tilde{x}\|^2 = f(\bar{x}, \tilde{x}) + \beta\sigma\|\bar{x} - \tilde{x}\|^2 = 0$$

and thus $\tilde{x} = \bar{x}$ follows since $\sigma < 1$. $\qquad\qquad\qquad\qquad\qquad\qquad\qquad$ $\square$

The chain of inequalities (3.9) shows that the next iterate is closer to the solution set of EP(f, C) than the current one. Hence, the hyperplane extragradient algorithm can be viewed as a descent method for the merit function that measures the distance from the solution set. Anyhow, since the solution set is unknown *a priori*, this merit function is not explicitly known and the classical descent techniques of the next section can not be exploited to minimize it.

3.2 Descent Methods

A descent method can be devised whenever a merit function is given in such a way that its value (and possibly its gradient) can be computed explicitly. For instance, this is the case of the gap and D-gap functions introduced in Sect. 2.2. According to Definition 2.2.3, consider any continuously differentiable merit function $\psi : \mathbb{R}^n \to \mathbb{R}$ on a closed and convex set X, that satisfies the following two properties:

(P1) the sublevel sets $\{x \in X : \psi(x) \leq c\}$ are bounded for any $c > 0$;
(P2) there exist two continuous functions $d : \mathbb{R}^n \to \mathbb{R}^n$ and $\sigma : \mathbb{R}^n \to \mathbb{R}$ such that any $x \in X$ not solving EP(f, C) satisfies $x + d(x) \in X$, $\sigma(x) > 0$ and

$$\langle \nabla\psi(x), d(x) \rangle \leq -\sigma(x).$$

Notice that *(P2)* implies that the stationary points of ψ over X (see Theorem 2.1.3) and its global minima coincide even though ψ is not necessarily convex.

Therefore, descent techniques can be exploited to minimize it just moving away form the current iterate x^k along the direction $d(x^k)$ with a step provided by a suitable line search rule.

Descent algorithm

Step 0. Choose $x^0 \in X$, $\delta, \eta \in (0, 1)$ and set $k = 0$.
Step 1. If $\psi(x^k) = 0$ then stop.
Step 2. Set $d^k = d(x^k)$, $\sigma_k = \sigma(x^k)$ and compute the smallest $s \in \mathbb{N}$ such that

$$\psi(x^k + \delta^s d^k) \le \psi(x^k) - \eta \delta^s \sigma_k$$

Step 3. Set $t_k = \delta^s$, $x^{k+1} = x^k + t_k d^k$, $k = k + 1$ and go to Step 1.

If the algorithm stops at Step 1, then the current iterate solves EP(f, C) since ψ is a merit function and the whole sequence $\{x^k\}$ lies in X thanks to the convexity of X.

Property *(P2)* guarantees that the line search procedure at Step 2 is finite. In fact, suppose *ab absurdo* that the inequality

$$\psi(x^k + \delta^s d^k) - \psi(x^k) > -\eta \delta^s \sigma_k$$

holds for all $s \in \mathbb{N}$. Taking the limit as $s \to +\infty$ leads to

$$\langle \nabla \psi(x^k), d^k \rangle \ge -\eta \sigma_k > -\sigma_k$$

that contradicts *(P2)* since x^k does not solve EP(f, C). Therefore, the value of the merit function is improved at each iteration and convergence can be proved in the following way.

Theorem 3.2.1 *If the descent algorithm produces an infinite sequence $\{x^k\}$, then it is bounded and any of its cluster points solves EP(f, C).*

Proof The sequence is bounded since $\{\psi(x^k)\}$ is monotone decreasing and the sublevel sets of ψ are bounded thanks to *(P1)*. Let $\bar{x}$ be any cluster point of the sequence and let $x^{k_\ell} \to \bar{x}$ for some $k_\ell \uparrow +\infty$ as $\ell \uparrow +\infty$. The continuity of d and σ guarantees $d^{k_\ell} = d(x^{k_\ell}) \to d(\bar{x}) = \bar{d}$ and $\sigma_{k_\ell} = \sigma(x^{k_\ell}) \to \sigma(\bar{x}) = \bar{\sigma}$.

By contradiction, suppose $\bar{x}$ does not solve EP(f, C) and hence $\bar{\sigma} > 0$ by *(P2)*. Since the sequence $\{\psi(x^k)\}$ is non-negative, it has a limit and thus

$$\lim_{\ell \to +\infty} \left[\psi(x^{k_\ell}) - \psi(x^{k_{\ell+1}}) \right] = 0$$

holds as well. Moreover, the stepsize rule guarantees

$$\psi(x^{k_\ell}) - \psi(x^{k_{\ell+1}}) \ge \psi(x^{k_\ell}) - \psi(x^{k_\ell+1}) \ge \eta \, t_{k_\ell} \, \sigma_{k_\ell} > 0.$$

Therefore, $t_{k_\ell} \to 0$ as $\ell \to +\infty$ since $\bar{\sigma} > 0$. On the other side, the inequality

$$\psi\left(x^{k_\ell} + t_{k_\ell}\,\delta^{-1}\,d^{k_\ell}\right) - \psi(x^{k_\ell}) > -\eta\left(t_{k_\ell}\,\delta^{-1}\right)\sigma_{k_\ell}$$

holds for all $\ell \in \mathbb{N}$ by the line search procedure, while the Mean Value Theorem A.2.6 guarantees the existence of some $\theta_\ell \in (0, 1)$ such that

$$\psi\left(x^{k_\ell} + t_{k_\ell}\,\delta^{-1}\,d^{k_\ell}\right) - \psi(x^{k_\ell}) = \left\langle \nabla\psi\left(x^{k_\ell} + \theta_\ell\,t_{k_\ell}\,\delta^{-1}\,d^{k_\ell}\right), t_{k_\ell}\,\delta^{-1}\,d^{k_\ell}\right\rangle.$$

Hence, the two inequalities together provide

$$\left\langle \nabla\psi\left(x^{k_\ell} + \theta_\ell\,t_{k_\ell}\,\delta^{-1}\,d^{k_\ell}\right), d^{k_\ell}\right\rangle > -\eta\,\sigma_{k_\ell}.$$

Taking the limit, the inequalities

$$\langle \nabla\psi(\bar{x}), \bar{d}\rangle \geq -\eta\,\bar{\sigma} > -\bar{\sigma},$$

contradict *(P2)*. Therefore, $\bar{x}$ solves EP(f, C). $\qquad\qquad\square$

Since the sequence $\{x^k\}$ is bounded, the above convergence theorem yields the existence of a solution of the Ky Fan inequality EP(f, C) as well.

It is worth noting that no assumptions are actually made on f in the above framework. Indeed, the tools for the convergence of the algorithm lie in the differentiability of ψ and the properties *(P1)* and *(P2)*. The next subsections show that the gap and D-gap functions introduced in Sect. 2.2 are concrete merit functions that satisfy *(P1)* and *(P2)* under suitable differentiability, convexity and monotonicity assumptions.

Gap Functions

Given any $\alpha > -\tau$, the gap function (2.17), i.e.,

$$\varphi_\alpha(x) = -\min\{\,f(x, y) + \alpha\,\|y - x\|^2/2 \;:\; y \in C\,\},$$

is a merit function on C and there exists a unique minimizer $y_\alpha(x)$ of the inner optimization problem (see Sect. 2.2). Moreover, suppose that f is continuously differentiable so that also φ_α is continuously differentiable provided that C is compact (see Theorem 2.2.7). The other properties that are needed to achieve convergence according to Theorem 3.2.1 hold, provided that suitable concavity conditions on f are met.

Theorem 3.2.2 *Suppose C is compact. Then, φ_α satisfies (P1). If moreover $f(\cdot, y)$ is γ-concave for all $y \in C$ for some $\gamma > -\tau$, then φ_α satisfies (P2) together with $d(x) = y_\alpha(x) - x$ and $\sigma(x) = (\gamma + \tau)\|y_\alpha(x) - x\|^2/2$.*

Proof (P1) holds since $X = C$ is bounded. The continuity of y_α (see Corollary 2.2.12) guarantees that also the functions d and σ are continuous. Any $x \in C$ not solving EP(f, C) satisfies $x + d(x) = y_\alpha(x) \in C$ and $\sigma(x) > 0$. Given any $y \in C$, the function $g(x) = f(x, y) + \gamma\|x\|^2/2$ is concave so that the relation $g(y) \leq g(x) + \langle \nabla g(x), y - x \rangle$ holds and it reads

$$\gamma\|y\|^2/2 \leq f(x, y) + \gamma\|x\|^2/2 + \langle \nabla_x f(x, y) + \gamma x, y - x \rangle$$

or equivalently

$$f(x, y) + \langle \nabla_x f(x, y), y - x \rangle \geq \gamma\|y - x\|^2/2. \tag{3.11}$$

Therefore, (P2) follows from the following chain of equality and inequalities

$$
\begin{aligned}
\langle \nabla\varphi_\alpha(x), y_\alpha(x) - x \rangle &= -\langle \nabla_x f(x, y_\alpha(x)), y_\alpha(x) - x \rangle + \alpha\|y_\alpha(x) - x\|^2 \\
&\leq f(x, y_\alpha(x)) - \gamma\|y_\alpha(x) - x\|^2/2 + \alpha\|y_\alpha(x) - x\|^2 \\
&\leq \langle \nabla_y f(x, y_\alpha(x)), y_\alpha(x) - x \rangle \\
&\quad -(\gamma + \tau)\|y_\alpha(x) - x\|^2/2 + \alpha\|y_\alpha(x) - x\|^2 \\
&\leq -(\gamma + \tau)\|y_\alpha(x) - x\|^2/2,
\end{aligned}
$$

where the first inequality is due to (3.11), the second to the τ-convexity and the third follows from the first-order optimality condition for $y_\alpha(x)$. □

It is worth noting that the theorem still holds replacing the concavity assumption with the slightly weaker condition (3.11). When EP(f, C) is a variational inequality, i.e., $f(x, y) = \langle F(x), y - x \rangle$ for some continuously differentiable map $F : \mathbb{R}^n \to \mathbb{R}^n$, condition (3.11) reads

$$\langle y - x, \nabla F(x)(y - x) \rangle \geq \gamma\|y - x\|^2/2.$$

If this inequality holds for any $x, y \in \mathbb{R}^n$, it actually states that all the matrices $\nabla F(x) - \mu I$ with $\mu = \gamma/2$ are positive semidefinite, that in turn is equivalent to the μ-monotonicity of F (see Sect. 2.1).

D-Gap Functions

Given any α and β with $\beta > \alpha > -\tau$, the D-gap function (2.19), i.e.,

$$\varphi_{\alpha\beta}(x) = \varphi_\alpha(x) - \varphi_\beta(x),$$

is a merit function on the whole space $\mathbb{R}^n$ (see Sect. 2.2). In this way, techniques for unconstrained optimization can be exploited to solve Ky Fan inequalities. Anyhow, each computation of the D-gap function amounts to solving two strongly convex optimization problems, while just one is needed for the gap function in the constrained case of the previous subsection. Moreover, suppose that f is continuously differentiable so that also $\varphi_{\alpha\beta}$ is continuously differentiable provided that C is compact (see Theorem 2.2.13).

In order to get a descent direction, the joint exploitation of φ_α and φ_β suggests to rely on the direction

$$r(x) = y_\alpha(x) - y_\beta(x)$$

paired with the additional term

$$s(x) = \alpha(x - y_\alpha(x)) - \beta(x - y_\beta(x)),$$

which is provided by the gradient of $\varphi_{\alpha\beta}$. The properties that are needed to achieve convergence according to Theorem 3.2.1 hold, provided that suitable monotonicity and Lipschitz conditions on partial gradients of f are met.

Theorem 3.2.3 *Suppose C is compact. Then, $\varphi_{\alpha\beta}$ satisfies (P1). If moreover $\nabla_x f(x, \cdot)$ is Lipschitz continuous with constant $L > 0$ and μ-monotone for some $\mu > -\tau$ for all $x \in \mathbb{R}^n$ and*

$$0 < \rho \le \min\left\{1/(\mu+\tau), \ (\mu+\tau)/(\mu+\tau+L)^2\right\},$$

then $\varphi_{\alpha\beta}$ satisfies (P2) together with $d(x) = r(x) + \rho s(x)$ and $\sigma(x) = (\mu + \tau)(\|r(x)\| + \rho\|s(x)\|)^2/2$.

Proof By contradiction, suppose that *(P1)* does not hold, i.e. there exist $c > 0$ and an unbounded sequence $\{x^k\}$ such that $\varphi_{\alpha\beta}(x^k) \le c$ for any $k \in \mathbb{N}$. On the other hand, the D-gap function satisfies $\varphi_{\alpha\beta}(x^k) \ge (\beta - \alpha)\|x^k - y_\beta(x^k)\|^2/2$ (see Theorem 2.2.13). The unboundedness of the sequence $\{x^k\}$ paired with the boundedness of the sequence $\{y_\beta(x^k)\}$, that belongs to C, leads to the contradiction

$$c \ge \varphi_{\alpha\beta}(x^k) \ge (\beta - \alpha)\|x^k - y_\beta(x^k)\|^2/2 \to +\infty.$$

The functions d and σ are continuous thanks to the continuity of y_α (see Corollary 2.2.12) and $\sigma(x) > 0$ holds for any $x \in \mathbb{R}^n$ not solving EP(f, C) by Theorem 2.2.13. Moreover, the first-order optimality conditions guarantee that $y_\alpha(x)$ and $y_\beta(x)$ satisfy

$$\langle \nabla_y f(x, y_\alpha(x)) + \alpha(y_\alpha(x) - x), y_\beta(x) - y_\alpha(x)\rangle \ge 0,$$
$$\langle \nabla_y f(x, y_\beta(x)) + \beta(y_\beta(x) - x), y_\alpha(x) - y_\beta(x)\rangle \ge 0,$$

so that summing up the two inequalities provides

$$\langle \nabla_y f(x, y_\beta(x)) - \nabla_y f(x, y_\alpha(x)) + s(x), r(x) \rangle \geq 0.$$

The τ-convexity of $f(x, \cdot)$ implies that the map $\nabla_y f(x, \cdot)$ is τ-monotone (see Sect. 2.1) and thus the above inequality guarantees also

$$\langle s(x), r(x) \rangle \geq \langle \nabla_y f(x, y_\alpha(x)) - \nabla_y f(x, y_\beta(x)), r(x) \rangle \geq \tau \|r(x)\|^2. \tag{3.12}$$

Therefore, the following chain of equalities and inequalities holds

$$
\begin{aligned}
\langle \nabla \varphi_{\alpha\beta}(x), d(x) \rangle &= \langle \nabla_x f(x, y_\beta(x)) - \nabla_x f(x, y_\alpha(x)), r(x) \rangle \\
&\quad + \rho \langle \nabla_x f(x, y_\beta(x)) - \nabla_x f(x, y_\alpha(x)), s(x) \rangle \\
&\quad - \langle s(x), r(x) \rangle - \rho \|s(x)\|^2 \\
&\leq -\mu \|r(x)\|^2 + \rho \langle \nabla_x f(x, y_\beta(x)) - \nabla_x f(x, y_\alpha(x)), s(x) \rangle \\
&\quad - \langle s(x), r(x) \rangle - \rho \|s(x)\|^2 \\
&\leq -(\mu + \tau) \|r(x)\|^2 \\
&\quad + \rho \langle \nabla_x f(x, y_\beta(x)) - \nabla_x f(x, y_\alpha(x)), s(x) \rangle - \rho \|s(x)\|^2 \\
&\leq -(\mu + \tau) \|r(x)\|^2 + \rho L \|r(x)\| \|s(x)\| - \rho \|s(x)\|^2 \\
&= -(\mu + \tau)(\|r(x)\| + \rho \|s(x)\|)^2 / 2 \\
&\quad - (\sqrt{\mu + \tau} \|r(x)\| - \sqrt{\rho} \|s(x)\|)^2 / 2 \\
&\quad + \rho[(\mu + \tau)\rho - 1] \|s(x)\|^2 / 2 \\
&\quad + \sqrt{\rho} \left[\sqrt{\rho}(\mu + \tau + L) - \sqrt{\mu + \tau} \right] \|r(x)\| \|s(x)\| \\
&\leq -(\mu + \tau)(\|r(x)\| + \rho \|s(x)\|)^2 / 2,
\end{aligned}
$$

where the first inequality is due to monotonicity of $\nabla_x f(x, \cdot)$, the third to its Lipschitz continuity, the second to (3.12) and the last to the choice of ρ. □

Notice that if $\nabla_x f(x, \cdot)$ is μ-monotone, then (3.11) holds with $\gamma = 2\mu + \tau$. Indeed, μ-monotonicity guarantees

$$
\begin{aligned}
f(x, y) + \langle \nabla_x f(x, y), y - x \rangle &\geq f(x, y) + \langle \nabla_x f(x, x), y - x \rangle + \mu \|y - x\|^2 \\
&= f(x, y) - \langle \nabla_y f(x, x), y - x \rangle + \mu \|y - x\|^2 \\
&\geq (2\mu + \tau) \|y - x\|^2 / 2,
\end{aligned}
$$

where the equality holds since $\nabla_x f(x, x) = -\nabla_y f(x, x)$ follows immediately from $f(x, x) = 0$ for all $x \in \mathbb{R}^n$ while the last inequality is due to the convexity. Notice also that $\mu > -\tau$ holds if and only if $\gamma = 2\mu + \tau > -\tau$. Therefore, the assumptions of Theorem 3.2.3 guarantee also that the gap function φ_α satisfies $(P1)$ and $(P2)$.

When $EP(f, C)$ is a variational inequality, i.e., $f(x, y) = \langle F(x), y - x \rangle$ for some continuously differentiable map $F : \mathbb{R}^n \to \mathbb{R}^n$, then

$$\nabla_x f(x, z) - \nabla_x f(x, y) = \nabla F(x)(z - y)$$

so that the μ-monotonicity of $\nabla_x f(x, \cdot)$ amounts to the positive semidefiniteness of all the matrices $\nabla F(x) - \mu I$, that is the μ-monotonicity of F (see Sect. 2.1). Hence, as $\tau = 0$ holds for variational inequalities, the strong monotonicity of F guarantees the convergence of the descent algorithm both with the gap and the D-gap function.

3.3   Regularization Methods

Unlike the methods of the previous sections, the regularization methods do not exploit a reformulation of the Ky Fan inequality $EP(f, C)$.

Indeed, the key idea is to rely on a sequence of Ky Fan inequalities in such a way their solutions converge to a solution of $EP(f, C)$. Clearly, these inequalities have to be easier to solve than the original one. In particular, the addition of suitable regularization terms to f allows achieving stronger monotonicity or concavity properties (see Sect. 3.4 for details) so that the algorithms of the previous sections can be successfully applied.

Two different approaches are presented. The first (Tikhonov–Browder) always exploits the same regularization term driven by a parameter going to zero, while the term of the proximal point approach depends on the previous iterate and a fixed positive parameter.

Tikhonov–Browder algorithm

Step 0.   Choose a sequence $\{\varepsilon_k\} \downarrow 0$ and set $k = 0$.

Step 1.   Compute a solution x^k of $EP(g_k, C)$ with

$$g_k(x, y) = f(x, y) + \varepsilon_k \langle x, y - x \rangle.$$

Step 2.   Set $k = k + 1$ and go to Step 1.

Notice that no theoretical stopping criterion is available since the regularization parameter needs to go to zero to achieve convergence.

Theorem 3.3.1 *Suppose C is compact. Then, any cluster point of the bounded sequence $\{x^k\}$ produced by the Tikhonov–Browder algorithm is a solution of $EP(f, C)$. If moreover f is pseudomonotone, then the whole sequence converges to the least-norm solution of $EP(f, C)$.*

Proof Theorem 2.3.4 guarantees that each Ky Fan inequality $EP(g_k, C)$ admits at least one solution. Thus, the sequence $\{x^k\}$ is well defined and bounded since C is bounded. Let $\tilde{x}$ be any of its cluster point and consider the suitable subsequence $x^{k_\ell} \to \tilde{x}$. Since x^{k_ℓ} solves $EP(g_{k_\ell}, C)$, that is any $y \in C$ satisfies

$$f(x^{k_\ell}, y) \geq -\varepsilon_{k_\ell} \langle x^{k_\ell}, y - x^{k_\ell} \rangle,$$

$f(\tilde{x}, y) \geq 0$ follows taking the limit in the above inequality. Hence, $\tilde{x}$ solves $EP(f, C)$.

If f is pseudomonotone, then the solution set S of $EP(f, C)$ is closed and convex (see Theorem 2.2.15). Therefore, the least-norm solution $\bar{x}$ is the projection of 0 onto S, i.e., $\bar{x} = P_S(0)$ (see Theorem 2.1.4). As x^k solves $EP(g_k, C)$, then

$$0 \leq f(x^k, \bar{x}) + \varepsilon_k \langle x^k, \bar{x} - x^k \rangle \leq \varepsilon_k \langle x^k, \bar{x} - x^k \rangle$$

follows since $f(\bar{x}, x^k) \geq 0$ guarantees also $f(x^k, \bar{x}) \leq 0$ by the pseudomonotonicity of f. The above inequalities imply $\|x^k\| \leq \|\bar{x}\|$ since $\langle x^k, \bar{x} \rangle \leq (\|x^k\|^2 + \|\bar{x}\|^2)/2$. Therefore, any cluster point $\tilde{x}$ of $\{x^k\}$ satisfies $\|\tilde{x}\| \leq \|\bar{x}\|$. The uniqueness of the projection guarantees $\tilde{x} = \bar{x}$ since $\tilde{x}$ solves $EP(f, C)$ as well. Therefore, whole sequence $\{x^k\}$ converges to $\bar{x}$. $\qquad\square$

Remark 3.3.2 Theorem 3.3.1 still holds if the boundedness of C is replaced by the coercivity condition (2.24). In fact, it is possible to prove that the bifunction g_k satisfies (2.24) as well. Therefore, Theorem 2.3.5 guarantees that the existence of at least one solution of $EP(g_k, C)$, so that the sequence $\{x^k\}$ is well defined. Moreover, it is possible to prove that $\{x^k\}$ is bounded and hence the rest of the proof is exactly the same as above.

Since ε_k is bounded to go to zero, the difference between f and g_k gets smaller and smaller so that the effect of the regularization vanishes. In order to overcome this issue, a different regularization term can be considered. Anyhow, assumptions have to be strengthened to achieve convergence: the proximal point algorithm requires the existence of a solution that solves the Minty inequality as well, while the Tikhonov–Browder algorithm requires just the existence of a solution with no further properties.

Proximal point algorithm

Step 0. Choose $x^0 \in C$, $\vartheta > 0$ and set $k = 1$.

Step 1. Compute a solution x^k of $EP(h_k, C)$ with

$$h_k(x, y) = f(x, y) + \vartheta \langle x - x^{k-1}, y - x \rangle.$$

Step 2. If $x^k = x^{k-1}$, then stop.

Step 3. Set $k = k + 1$ and go to Step 1.

If the algorithm stops at step 2, clearly x^k solves $EP(f, C)$ since h_k and f coincide as the regularization term vanishes.

Theorem 3.3.3 *Suppose C is compact and the Minty inequality MEP(f, C) admits at least one solution. If the proximal point algorithm produces an infinite sequence $\{x^k\}$, then any of its cluster points is a solution of EP(f, C). If moreover f is pseudomonotone, then the whole sequence converges to one solution of EP(f, C).*

Proof Theorem 2.3.4 guarantees that each Ky Fan inequality $EP(h_k, C)$ admits at least one solution. Thus, the sequence $\{x^k\}$ is well defined and bounded since C is bounded. Let $\hat{x}$ be one solution of $MEP(f, C)$. Then, the following chain of equalities and inequalities holds

$$\|x^k - \hat{x}\|^2 = \|x^k - x^{k+1} + x^{k+1} - \hat{x}\|^2$$
$$= \|x^k - x^{k+1}\|^2 + \|x^{k+1} - \hat{x}\|^2 + 2\langle x^k - x^{k+1}, x^{k+1} - \hat{x}\rangle$$
$$\geq \|x^k - x^{k+1}\|^2 + \|x^{k+1} - \hat{x}\|^2 - 2f(x^{k+1}, \hat{x})/\vartheta$$
$$\geq \|x^k - x^{k+1}\|^2 + \|x^{k+1} - \hat{x}\|^2$$
$$\geq \|x^{k+1} - \hat{x}\|^2,$$

where the first inequality comes from x^k solving $EP(h_k, C)$ and the second from $\hat{x}$ solving $MEP(f, C)$. Therefore, $\{\|x^k - \hat{x}\|\}$ is monotone non-increasing sequence and thus convergent. As a consequence, the above chain guarantees also $\|x^k - x^{k+1}\| \to 0$. Let $\tilde{x}$ be any cluster point of $\{x^k\}$ and consider the suitable subsequence $x^{k_\ell} \to \tilde{x}$. Since x^{k_ℓ} solves $EP(h_{k_\ell}, C)$, that is any $y \in C$ satisfies

$$f(x^{k_\ell}, y) \geq -\vartheta\langle x^{k_\ell} - x^{k_\ell-1}, y - x^{k_\ell}\rangle,$$

$f(\tilde{x}, y) \geq 0$ follows taking the limit in the above inequality. Hence, $\tilde{x}$ solves $EP(f, C)$.

If f is also pseudomonotone, then $\tilde{x}$ solves $MEP(f, C)$ as well. Choosing $\hat{x} = \tilde{x}$ the limit of the sequence $\{\|x^k - \hat{x}\|\}$ satisfies

$$\lim_{k \to +\infty} \|x^k - \tilde{x}\| = \lim_{\ell \to +\infty} \|x^{k_\ell} - \tilde{x}\| = 0,$$

i.e., the whole sequence $\{x^k\}$ converges to $\tilde{x}$. □

Notice that the boundedness of C and the convexity of $f(x, \cdot)$ are needed just to prove the existence of one solution of $EP(h_k, C)$ and they do not play any role in the actual proof of convergence. Thus, the theorem could be stated replacing boundedness and convexity by the latter condition. Anyhow, convexity indirectly plays another role as it is needed to solve $EP(h_k, C)$ exploiting, for instance, one of the algorithms of the previous sections.

3.4   Computational Issues

Stopping Criteria

The stopping criteria in the algorithms are theoretical, in the sense that they are given via sufficient conditions for the current iterate to be a solution. Clearly, it is very unlikely that they are satisfied numerically so that some kind of approximation has to be devised.

A very small value of a merit function is not necessarily a good measure of the quality of the solution since merit functions are not scale free, that is any positive multiple of a merit function is still a merit function.

On the contrary, error bounds in Sect. 2.5 are scale free since merit functions are coupled with suitable constants and indeed they provide an estimate of the distance of the point from the solution. This works for the fixed-point and descent algorithms. In fact, the assumptions that guarantee their convergence allow exploiting one of the error bounds: Theorem 3.1.1 can be paired with Theorem 2.5.2, Theorem 3.1.2 with Theorem 2.5.3 and so on.

Differently, no error bound is available for the extragradient and proximal point algorithms except for the pseudomonotone case whenever $\tau > 0$ (Theorem 2.5.3). Clearly, the stopping criteria of steps 2 can be naturally approximated by the distance between the two points, hence the algorithms stop when it goes below some given threshold, but no estimate of the distance from the actual solution is guaranteed.

Both the regularization methods need to solve an equilibrium (sub)problem at each iteration, for instance exploiting the other algorithms of the chapter and their stopping criteria according to monotonicity or concavity assumptions that are satisfied. Notice that the monotonicity of the regularized bifunctions g_k and h_k is stronger than that of f since they are $(\mu + \varepsilon_k)$ and $(\mu + \vartheta)$-monotone when f is μ-monotone. Similarly, the γ-concavity of $f(\cdot, y)$ becomes $(\gamma + 2\varepsilon_k)$ and $(\gamma + 2\vartheta)$-concavity of $g_k(\cdot, y)$ and $h_k(\cdot, y)$, respectively, while the τ- convexity of $f(x, \cdot)$ does not change for $g_k(x, \cdot)$ and $h_k(x, \cdot)$.

Actually, also the single iterations of the fixed-point, extragradient and descent algorithms require some kind of approximation since they amount to the solution of convex optimization problems. Overall, it is worth developing inexact variants of all the algorithms of the chapter (see Sect. 3.5).

Parameters' Calibration

All the solution methods involve some parameters that have to be chosen *a priori*. The fixed point and extragradient algorithms require the choice of parameter α for the auxiliary Ky Fan inequality, the hyperplane extragradient and descent algorithms the choice of some parameters for the linesearch procedures, while the Tikhonov–Browder and proximal point algorithms the choice of regularization sequence $\{\varepsilon_k\}$ and parameter ϑ.

Generally, the choice of the parameters has to be performed numerically through suitable calibration procedures. Anyhow, whenever a rate of convergence is available, it is possible to analytically compute the parameter α that provides the best rate. This always happens for the fixed point algorithm (Theorems 3.1.1 and 3.1.2) and in one case for the extragradient algorithm (Theorem 3.1.3 c). In all the three cases the best parameter is the smallest possible value according to the assumptions of the convergence theorems.

However, this best parameter depends on the knowledge of the constants T_1, T_2 and τ. Indeed, the convergence results for the fixed point, extragradient, descent and (indirectly) regularization algorithms require the explicit knowledge of the constants of convexity, monotonicity, concavity and Lipschitz-type properties of the equilibrium bifunction f. Unless particular cases are at hand (such as affine Ky Fan inequalities, that is whenever $f(x, y) = \langle Px + Qy + r, y - x \rangle$ with P, Q being $n \times n$ matrices and $r \in \mathbb{R}^n$), estimates of these constants are difficult to get and parameters have to be chosen with no guarantee of actual convergence. To overcome these troubles, more complex algorithms may be devised by changing the parameters iteration by iteration in such a way that the knowledge of constants is no longer needed (see Sect. 3.5).

3.5 Notes and References

The algorithms of this chapter have been selected to represent the different classes of available algorithms, one for each class, choosing possibly the most basic ones. More advanced algorithms are available and pointers to literature are given below.

The fixed point approach for solving Ky Fan inequalities has been originally considered in [91] and its rate of convergence proved in [100] for the convex case under strong monotonicity assumptions. Theorem 3.1.1 subsumes the convergence results of these papers adding the analysis of the strongly convex case ($\tau > 0$) as well. A fairly new convergence result involving pseudomonotonicity is given in Theorem 3.1.2 exploiting similar techniques. Actually, the overall assumptions of the theorem still guarantee the uniqueness of a solution. In order to achieve convergence without uniqueness, a more complex approach has been developed in [103]. Considering suitable averages of all the iterates, other versions of the fixed point algorithm have been developed to improve the rate of convergence [112] and achieve convergence under monotonicity [10].

The extragradient approach has been originally considered in [67] and further analysed in [87, 88] under a Lipschitz-type assumption different from (3.1). The convergence results of Theorem 3.1.3 rely on [113], anyhow the analysis of the strongly convex case is new and provides also a rate of convergence which has not been highlighted in the literature. Replacing the regularization term with a suitable barrier function allows performing unconstrained minimization during the iterations [6, 7, 104]. A two-step extragradient method, which employs two fixed point type steps, has been developed to improve the numerical behaviour of the algorithm [102]. The effect of inexact resolution of the optimization subproblems is analysed in [67, 87, 88].

The hyperplane extragradient algorithm has been proposed in [113] by replacing the correction step of the original extragradient algorithm with a double projection, the first on a suitable hyperplane and the second on the feasible region. This basic version is reported in Sect. 3.1, while more advanced versions require the projection on their intersection [8, 9, 59] or the projection on the intersection with all the previously determined hyperplanes [58, 125]. Another kind of predictive step, which relies on all the previous iterates and one projection, has been considered in [114].

The general scheme of the descent algorithm given in Sect. 3.2 has been proposed in [34] together with its convergence proof. This scheme actually requires the uniqueness of the solution. Convergence can be achieved without uniqueness under weaker assumptions by exploiting a family of merit functions, for instance changing the parameters α and β in gap and D-gap functions [28, 29, 36].

The version of the general scheme with the gap function has been originally considered with the exact line search in [92] and with inexact line searches in [50, 82]. For the sake of simplicity the assumptions of Theorem 3.2.2 are slightly stronger than needed. Anyhow, also the assumptions in those papers guarantee that any stationary point of the gap function is actually a global minimum. In order to achieve convergence when this is not necessarily true, a more complex approach based on descent techniques has been developed in [36]. Since the evaluation of the gap function may involve convex optimization with nonlinear constraints, linearization procedures have been exploited together with penalization techniques in [28, 30]. Inexact resolution of the optimization subproblems paired with a suitable non-monotone line search is analysed in [57].

The version with the D-gap function has been originally considered in [52, 83] for the convex case, focusing on the descent property ($P2$). The boundedness of the sublevel sets has been originally proved in [135] supposing the boundedness of the feasible region. Actually, it holds also under monotonicity and Lipschitz properties of suitable gradient maps [52, 83, 136]. Some of these results are subsumed in Theorem 3.2.3 that in addition provides also the analysis of the strongly convex case. Another descent technique that pairs this approach together with classical gradient steps has been considered in [136, 137]. Just like in the case of gap functions, the assumptions of all these papers guarantee that stationary points coincide with global minima. In order to achieve convergence without this restrictive feature, a more complex descent approach has been developed in [29].

The Tikhonov–Browder algorithm has been proposed in [98]. Convergence can be achieved replacing the boundedness of the feasible region with suitable coercivity conditions [79, 81]. Since an exact solution of the regularized problem is difficult to compute in practice, inexact versions have been developed in [80, 84]. More general regularization terms have been analyzed as well [2, 72].

The analysis of the convergence of the proximal point algorithm given in Theorem 3.3.3 relies on both [97] and [78]. The former provides the analysis of the exact version under monotonicity assumption, while the latter exploits the same assumptions of Theorem 3.3.3 to analyze an inexact version of the algorithm. Other inexact versions have been proposed in [60, 97, 119]. More general regularization terms have been developed in [42].

Another way to solve a Ky Fan inequality relies on its equivalent reformulation as the variational inequality (2.15) and to apply solution methods for this latter problem. Projection type methods have been devised in [11, 20, 118, 120], while extragradient algorithms combined with hyperplane techniques similar to the one in Sect. 3.1 have been developed in [70, 77, 85, 132].

Whenever the Ky Fan inequality is equivalent to the Minty inequality (see Theorem 2.2.15), it reduces to the so-called convex feasibility problem [19], that is finding a point in the intersection of a family of convex sets. Therefore, any method for this latter problem could be exploited. In particular, *ad hoc* methods based on projection [73] and cutting plane techniques [115] have been developed.

Chapter 4
Quasi-Equilibria

A quasi-equilibrium problem is a Ky Fan inequality whose feasible region is subject to modifications according to the point considered as a candidate solution. Variable feasible regions are well suited to model situations in which players or (economic) agents share resources or more generally their supposed behaviour may influence also the feasible behaviours of the others.

A multivalued map $C : \mathbb{R}^n \rightrightarrows \mathbb{R}^n$ is required for describing how the feasible region changes together with the considered point and the corresponding mathematical model reads

$$\text{find } \bar{x} \in C(\bar{x}) \text{ such that } f(\bar{x}, y) \geq 0 \text{ for all } y \in C(\bar{x}) \qquad \text{QEP}(f, C)$$

where $f : \mathbb{R}^n \times \mathbb{R}^n \to \mathbb{R}$ is an equilibrium bifunction, i.e., it satisfies $f(x, x) = 0$ for any $x \in \mathbb{R}^n$. The reformulation of the above quasi-equilibrium problem as an optimization problem, some basic existence results and algorithms are analysed along the same lines of Chaps. 2 and 3 for Ky Fan inequalities.

4.1 Applications

Some reasonable features of the problems described in Chap. 1 have not been considered in order to keep the corresponding models as simple as possible. For instance, it is reasonable that the users of a wireless system seek at least some minimal quality of the signal, that depends upon the interferences due to the transmit power of all the users. Likewise, some users of a road traffic network may give up a plan of travel if they believe the network to be too crowded: hence, the traffic demand may be influenced by the perceived state of the network, that depends upon the supposed behaviour of all the users. Variable feasible regions can be exploited to model such situations.

© Springer Nature Switzerland AG 2019
G. Bigi et al., *Nonlinear Programming Techniques for Equilibria*, EURO Advanced
Tutorials on Operational Research, https://doi.org/10.1007/978-3-030-00205-3_4

Power Control in Wireless Communications and Generalized Nash Equilibria

In the power control problem of Sect. 1.2 mobile users have to choose their power to minimize both the cost and the interference with the other devices. Anyway, the utility function c_i of user i given by (1.4) does not guarantee any minimum quality of service Q_i. Therefore, a constraint such as $\gamma_i(x) \geq Q_i$ can be added. In this way, the set of available strategies of the user depends on the strategies chosen by all the other users. Hence, it can be described by a multivalued map $C_i : \mathbb{R}^{N-1} \rightrightarrows \mathbb{R}$.

More in general, adopting the same notations of Sect. 1.8, the generalized Nash equilibrium problem involves N players, each controlling the variable $x_i \in \mathbb{R}^{n_i}$ in order to minimize its cost function $c_i : \mathbb{R}^n \to \mathbb{R}$ where $n = n_1 + \cdots + n_N$. The possible strategies of each player belong to a set described by a multivalued map $C_i : \mathbb{R}^{n-n_i} \rightrightarrows \mathbb{R}^{n_i}$, so that the strategy set $C_i(x_{-i})$ depends on the other players' strategies x_{-i}. A generalized Nash equilibrium is any vector $\bar{x} \in \mathbb{R}^n$ such that

$$\bar{x}_i \in \arg\min\{c_i(y_i, \bar{x}_{-i}) : y_i \in C_i(\bar{x}_{-i})\}$$

for any player i. The Nikaido–Isoda bifunction f given by (1.14) allows formulating the *generalized Nash equilibrium problem* as the quasi-equilibrium problem QEP(f, C) with $C(x) = C_1(x_{-1}) \times \cdots \times C_N(x_{-N})$ for any $x \in \mathbb{R}^n$.

Traffic Networks and Quasi-Variational Inequalities

In the traffic network problem of Sect. 1.3 the demand d_s for each origin-destination pair s is supposed to be a given rate. Anyway, this choice does not allow considering possible phenomena of elasticity of the demand, which may occur in dynamic settings when information on the state of the network is available. If the demands depend upon the minimum travel times, they ultimately depend upon the flow on the network. Therefore, at a steady state any feasible path flow x has to be consistent with the corresponding demands $d_s(x)$, that is it has to be a fixed point of the constraining map

$$C(x) = \left\{ y \in \mathbb{R}^n_+ : \sum_{p \in P_s} y_p = d_s(x), \quad \forall s \in OD \right\}.$$

According to Wardrop equilibrium principle, traffic flows only on paths with minimum travel time. Similarly to Sect. 1.3, it can be proved that a path flow $\bar{x}$ is a Wardrop equilibrium if and only if

$$\bar{x} \in C(\bar{x}) \text{ and } \langle T(\bar{x}), y - \bar{x} \rangle \geq 0 \text{ for all } y \in C(\bar{x}). \tag{4.1}$$

Problems like (4.1) are known as *quasi-variational inequalities* and they are the particular cases of QEP(f, C) where $f(x, \cdot)$ is affine for any $x \in \mathbb{R}^n$.

Portfolio Selection and Pareto Optima

The Markowitz portfolio selection problem of Sect. 1.4 aims at maximizing the expected return $g_1(x)$ of the chosen portfolio x while minimizing the corresponding risk, that is measured as the expected variance $g_2(x)$ of the portfolio. As the two goals are generally conflicting, non-dominated portfolios are considered reasonable selections. Ky Fan inequalities allow identifying portfolios x that are not strictly dominated (see Sect. 1.8). This is not completely satisfactory as two such portfolios may have the same expected return but different risk or vice versa. On the contrary, non-dominated portfolios do not have this drawback. Indeed, a portfolio x is non-dominated if no other portfolio x' satisfies both $g_1(x') \geq g_1(x)$ and $g_2(x') \leq g_2(x)$ with at least one of the two inequalities holding strictly. Such portfolios are called Pareto optimal.

More in general, adopting the same notations of Sect. 1.8, $\bar{x} \in X$ is a Pareto minimum of the vector function $\psi : \mathbb{R}^n \to \mathbb{R}^m$ over X if there exists no $y \in X$ such that $\psi_i(y) \leq \psi_i(\bar{x})$ holds for any $i = 1, \ldots, m$ and there exists an index j such that $\psi_j(y) < \psi_j(\bar{x})$. A variable feasible region helps identifying Pareto minima: consider

$$C(x) = \{y \in X : \psi_i(y) \leq \psi_i(x), \ i = 1, \ldots, m\}$$

and the bifunction

$$f(x, y) = \sum_{i=1}^{m} [\psi_i(y) - \psi_i(x)].$$

Therefore, finding a Pareto minimum of ψ over X amounts to solving QEP(f, C). Indeed, $\bar{x} \in C(\bar{x})$ if and only if $\bar{x} \in X$ while any $y \in C(\bar{x})$ satisfies $f(\bar{x}, y) \geq 0$ if and only if $\psi_i(y) = \psi_i(\bar{x})$ holds for any i.

4.2 Theory

The set of the fixed points of C, namely,

$$\text{fix } C = \{x \in \mathbb{R}^n : x \in C(x)\},$$

has to be nonempty for the quasi-equilibrium problem QEP(f, C) to have solutions. In fact, any of its solution must belong to fix C, that can be somehow considered as the feasible region of QEP(f, C). Actually, its reformulation as an optimization program through a merit function exploits fix C exactly in this way. Similarly to Sect. 2.2, consider the value function

$$\varphi_\alpha(x) = \max\{-f_\alpha(x, y) : y \in C(x)\} = -\min\{f_\alpha(x, y) : y \in C(x)\} \qquad (4.2)$$

where f_α is given by (2.16). The only difference with the gap function (2.17) for Ky Fan inequalities lies in the dependence of the feasible region in (4.2) upon the considered point. If $x \in$ fix C, then $\varphi_\alpha(x) \geq -f_\alpha(x, x) = 0$. Moreover, if $f(x, \cdot)$ is τ-convex and $\alpha > -\tau$, then $f_\alpha(x, \cdot)$ is $(\tau + \alpha)$-convex with $\tau + \alpha > 0$ and the set of optimal solutions $Y_\alpha(x) = \arg\min\{f_\alpha(x, y) : y \in C(x)\}$ reduces to the singleton $\{y_\alpha(x)\}$. As $\bar{x}$ solves QEP(f, C) if and only if $\bar{x} \in$ fix C and it solves EP$(f, C(\bar{x}))$, the following equivalences are a straightforward consequence of Corollary 2.2.11.

Theorem 4.2.1 *Suppose* fix C *is nonempty and* $f(x, \cdot)$ *is* τ-*convex with* $\tau \geq 0$ *for any* $x \in$ fix C. *Given any* $\alpha > -\tau$, *the following statements are equivalent:*

a) $\bar{x}$ *solves QEP(f, C),*
b) $y_\alpha(\bar{x}) = \bar{x}$,
c) $\bar{x} \in$ fix C *and* $\varphi_\alpha(\bar{x}) = 0$.

Therefore, φ_α can be referred to as a merit function on fix C for QEP(f, C): Theorem 4.2.1 states that any solution of QEP(f, C) solves the minimization problem

$$\min\{\varphi_\alpha(x) : x \in \text{fix } C\}, \qquad (4.3)$$

that is actually equivalent to the quasi-equilibrium problem provided that the optimal value is zero. Anyway, (4.3) is generally hard to solve: fix C may have a complex structure and φ_α is neither convex nor differentiable, while at least continuity is guaranteed provided that both lower and upper semicontinuity of the multivalued map C hold beyond the compactness of all sets $C(x)$ and the continuity of f (see Theorem 2.2.6 for a comparison).

Theorem 4.2.2 *Suppose* C *is both upper and lower semicontinuous multivalued map with nonempty compact values,* $f(x, \cdot)$ *is* τ-*convex for any* $x \in C$ *with* $\tau \geq 0$ *and* f *is continuous. Given any* $\alpha > -\tau$, *both the functions* $y_\alpha(\cdot)$ *and* φ_α *are continuous.*

Proof By contradiction, suppose $y_\alpha(\cdot)$ is not continuous at some $x \in \mathbb{R}^n$. Thus, there exist a neighborhood U of $y_\alpha(x)$ and a sequence $x^k \to x$ such that $y_\alpha(x^k) \notin U$. Since C is upper semicontinuous with compact values and $y_\alpha(x^k) \in C(x^k)$, there exists $y \in C(x)$ and a subsequence $\{y_\alpha(x^{k_\ell})\}$ such that $y_\alpha(x^{k_\ell}) \to y$ as $\ell \to +\infty$ (see Theorem A.3.5). Clearly, $y \neq y_\alpha(x)$ and hence $f_\alpha(x, y_\alpha(x)) < f_\alpha(x, y)$ holds. Since C is lower semicontinuous and $x^{k_\ell} \to x$, Theorem A.3.6 can be applied to this sequence together with $y_\alpha(x) \in C(x)$. Hence, taking a subsequence if necessary, there exist $y^{k_\ell} \in C(x^{k_\ell})$ such that $y^{k_\ell} \to y_\alpha(x)$. The continuity of f_α guarantees

$$\lim_{\ell \to +\infty} f_\alpha(x^{k_\ell}, y^{k_\ell}) = f_\alpha(x, y_\alpha(x)) < f_\alpha(x, y) = \lim_{\ell \to +\infty} f_\alpha(x^{k_\ell}, y_\alpha(x^{k_\ell}))$$

so that

$$f_\alpha(x^{k_\ell}, y^{k_\ell}) < f_\alpha(x^{k_\ell}, y_\alpha(x^{k_\ell}))$$

holds whenever ℓ is large enough, contradicting that $y_\alpha(x^{k_\ell})$ is a minimizer of $f_\alpha(x^{k_\ell}, \cdot)$ over $C(x^{k_\ell})$.

Finally, the continuity of φ_α follows immediately since f_α is continuous (as f is) and $\varphi_\alpha(x) = -f_\alpha(x, y_\alpha(x))$. □

Existence results for QEP(f, C) can be developed similarly to Sect. 2.3 for Ky Fan inequalities under suitable quasiconvexity and semicontinuity conditions on f. Anyway, beyond the equiboundedness of all the sets $C(x)$, some further assumptions on the behaviour of the multivalued map C and the bifunction f are needed as well.

Theorem 4.2.3 *Suppose C is a lower semicontinuous with nonempty convex values, fix C is closed and there exists a compact convex set K such that $C(x) \subseteq K$ for any $x \in \mathbb{R}^n$. If $f(\cdot, y)$ is upper semicontinuous for any $y \in \mathbb{R}^n$ and $f(x, \cdot)$ is quasiconvex for any $x \in K$ and upper semicontinuous for any $x \in \partial_K$ fix C, then QEP(f, C) has at least one solution.*

Proof Theorem A.3.11 guarantees that fix C is not empty. If fix $C = K$, any solution of EP(f, K) solves QEP(f, C) as well and the existence of at least one solution of EP(f, K) is granted by Theorem 2.3.4.

Now, suppose that fix C is a proper subset of K. Hence, fix C can not be open in K since it is closed and K is convex (see Appendix A.1). As a consequence, both int_K fix $C \neq$ fix C and ∂_K fix $C =$ fix $C \setminus \text{int}_K$ fix $C \neq \emptyset$ hold.

Consider the multivalued map $F : \mathbb{R}^n \rightrightarrows \mathbb{R}^n$ given by

$$F(x) = \{y \in \mathbb{R}^n : f(x, y) < 0\}$$

and the piecewise multivalued map $G : K \rightrightarrows \mathbb{R}^n$ given by

$$G(x) = \begin{cases} F(x) \cap K & \text{if } x \in \text{int}_K \text{ fix } C, \\ F(x) \cap C(x) & \text{if } x \in \partial_K \text{ fix } C, \\ C(x) & \text{if } x \notin \text{fix } C. \end{cases}$$

Since $K \setminus \text{fix } C$ is open in K, G is lower semicontinuous on $K \setminus \text{fix } C$ as it coincides with the lower semicontinuous multivalued map C on this open set. The upper semicontinuity of $f(\cdot, y)$ implies that the set

$$F^{-1}(y) := \{x \in \mathbb{R}^n : f(x, y) < 0\}$$

is open for every $y \in \mathbb{R}^n$ so that $F \cap K : x \mapsto F(x) \cap K$ has open lower sections in K. Therefore, $F \cap K$ is lower semicontinuous on $\text{int}_K \text{ fix } C$ by Theorem A.3.8 and consequently G is lower semicontinuous on this open set as well.

Since $f(\cdot, y)$ is upper semicontinuous, then F has open lower sections. Hence, Theorem A.3.8 guarantees that it is lower semicontinuous. Moreover, it has open convex values since $f(x, \cdot)$ is quasiconvex and upper semicontinuous for any given $x \in \partial_K \text{ fix } C$. Hence, the Open Graph Theorem A.3.7 guarantees that gph F is open and consequently $F \cap C : x \mapsto F(x) \cap C(x)$ is lower semicontinuous on $\partial_K \text{ fix } C$ by Theorem A.3.9.

Given any $x \in \partial_K \text{ fix } C$, consider any open set $\Omega \subseteq \mathbb{R}^n$ such that $G(x) \cap \Omega \neq \emptyset$ or equivalently $F(x) \cap C(x) \cap \Omega \neq \emptyset$. Thus, $F(x) \cap K \cap \Omega \neq \emptyset$ holds as well and $C(x) \subseteq K$ implies $C(x) \cap \Omega \neq \emptyset$. Since $F \cap K$ and C are lower semicontinuous, there exists a neighborhood U' of x such that $F(z) \cap K \cap \Omega \neq \emptyset$ and $C(z) \cap \Omega \neq \emptyset$ hold for any $z \in U'$. Moreover, the lower semicontinuity of $F \cap C$ on $\partial_K \text{ fix } C$ provides another neighbourhood U'' of x such that $F(z) \cap C(z) \cap \Omega \neq \emptyset$ holds for any $z \in U'' \cap \partial_K \text{ fix } C$. Therefore any $z \in U \cap K$, where $U = U' \cap U''$ is another neighbourhood of x, satisfies $G(z) \cap \Omega \neq \emptyset$, showing that G is lower semicontinuous at x.

Summing up all the case, G is lower semicontinuous on its whole domain. Moreover, its range is a subset of K and it has convex values since $f(x, \cdot)$ is quasiconvex for any $x \in K$. Supposing $\text{dom } G = K$, all the assumptions of Theorem A.3.11 would be satisfied so that there would exist $x \in \text{fix } G$. If $x \notin \text{fix } C$, then $x \in G(x) = C(x)$ that is not possible. Anyway, if $x \in \text{fix } C$, then $x \in \text{fix } F$ would hold implying the contradiction $f(x, x) < 0$. Therefore, there exists $\bar{x} \in K$ such that $G(\bar{x}) = \emptyset$. Since C has nonempty values, then $\bar{x} \in \text{fix } C$ and two cases may occur. If $\bar{x} \in \partial_K \text{ fix } C$, then $F(\bar{x}) \cap C(\bar{x}) = \emptyset$, i.e., it solves QEP($f, C$). If $\bar{x} \in \text{int}_K \text{ fix } C$, then $F(\bar{x}) \cap K = \emptyset$, i.e., it solves the equilibrium problem EP(f, K) and thus QEP(f, C) as well. $\square$

The proof actually shows that at least one solution of QEP(f, C) belongs either to the boundary $\partial_K \text{ fix } C$ or to the set $\{x \in \text{int}_K \text{ fix } C : x \text{ solves EP}(f, K)\}$. In particular, if EP($f, K$) has no solution, at least one solution of QEP(f, C) lies on the boundary of fix C in K.

Whenever $C(x) = C$ holds for any $x \in \mathbb{R}^n$ for some convex compact set $C \subseteq \mathbb{R}^n$, i.e., the quasi-equilibrium problem QEP(f, C) turns out to be the Ky Fan inequality EP(f, C), Theorem 4.2.3 collapses to Theorem 2.3.4 up to considering upper semicontinuity of $f(\cdot, y)$ for any $y \in \mathbb{R}^n$ rather than any $y \in C$. Indeed, both $K = C$ and fix $C = C$ hold so that $\partial_C \text{ fix } C = \emptyset$ and the assumption on the upper semicontinuity of the functions $f(x, \cdot)$ is no longer needed.

Just like for Ky Fan inequalities (see Theorem 2.3.5), the compactness of K can be replaced by suitable coercivity conditions. Indeed, the equiboundedness assumption can be dropped if the following coercivity condition

$$\forall x \in \text{fix } C \text{ with } \|x\| > r, \ \exists y \in C(x) \text{ with } \|y\| < \|x\| \text{ and } f(x, y) \le 0 \qquad (4.4)$$

holds for some $r > 0$ such that any $x \in \mathbb{R}^n$ satisfies the anchorage condition

$$C(x) \cap B(0, r) \neq \emptyset. \qquad (4.5)$$

Theorem 4.2.4 *Let $K \subseteq \mathbb{R}^n$ be a closed and convex set satisfying $C(x) \subseteq K$ for any $x \in \mathbb{R}^n$. Suppose C is lower semicontinuous with nonempty convex values, fix C is closed, $f(x, \cdot)$ is convex for any $x \in K$ and $f(\cdot, y)$ is upper semicontinuous for any $y \in \mathbb{R}^n$. If there exists $r > 0$ such that (4.5) holds for any $x \in \mathbb{R}^n$ and the coercivity condition (4.4) is satisfied, then QEP(f, C) has at least one solution.*

Proof Given any $R > r$, consider the multivalued map $C_R(x) = C(x) \cap \bar{B}(0, R)$. Chosen any $x \in K$ and an open set $\Omega \subseteq \mathbb{R}^n$ such that there exists $y \in C_R(x) \cap \Omega$, the anchorage condition (4.5) guarantees the existence of $z \in C(x) \cap B(0, R)$: whenever $t > 0$ is small enough, $tz + (1 - t)y \in C(x) \cap B(0, R) \cap \Omega$ holds as well. Moreover, the multivalued map $x \mapsto C(x) \cap B(0, R)$ is lower semicontinuous by Theorem A.3.9 since the graph of the constant multivalued map $x \mapsto B(0, R)$ is open. Therefore, there exists a neighborhood U of x such that

$$\emptyset \neq C(x') \cap B(0, R) \cap \Omega \subseteq C_R(x') \cap \Omega$$

holds for any $x' \in U$. Hence, C_R is lower semicontinuous as well.

Clearly, $C_R(x) \subseteq \bar{B}(0, R) \cap K$ holds for any $x \in K$ and fix $C_R = \text{fix } C \cap \bar{B}(0, R) \cap K$ is closed so that Theorem 4.2.3 guarantees the existence of a solution $\bar{x}$ of QEP(f, C_R), that is $\bar{x} \in \text{fix } C \cap \bar{B}(0, R) \cap K$ and $f(\bar{x}, y) \ge 0$ for any $y \in C(\bar{x})$ with $\|y\| \le R$.

Consider any $y \in C(\bar{x})$ such that $\|y\| > R$.

If $\|\bar{x}\| < R$, then $(1 - t)\bar{x} + ty \in C_R(\bar{x})$ holds whenever $t > 0$ is small enough. Therefore, the convexity of $f(\bar{x}, \cdot)$ implies

$$0 \le f(\bar{x}, (1 - t)\bar{x} + ty) \le (1 - t)f(\bar{x}, \bar{x}) + tf(\bar{x}, y) = tf(\bar{x}, y)$$

and hence $f(\bar{x}, y) \ge 0$.

If $\|\bar{x}\| = R$, the coercivity condition (4.4) implies the existence of $\bar{y} \in C(\bar{x})$ with $\|\bar{y}\| < R$ such that $f(\bar{x}, \bar{y}) \le 0$. Hence, $f(\bar{x}, \bar{y}) = 0$ holds since $\bar{x}$ solves QEP(f, C_R). As $(1 - t)\bar{y} + ty \in C_R(\bar{x})$ holds whenever $t > 0$ is small enough, the convexity of $f(\bar{x}, \cdot)$ implies

$$0 \le f(\bar{x}, (1 - t)\bar{y} + ty) \le (1 - t)f(\bar{x}, \bar{y}) + tf(\bar{x}, y) = tf(\bar{x}, y)$$

and hence $f(\bar{x}, y) \geq 0$.

Summing up all the situations, $\bar{x}$ solves QEP(f, C) as well.   □

Whenever $C(x) = C$ holds for any $x \in \mathbb{R}^n$ for some closed convex set $C \subseteq \mathbb{R}^n$, the anchorage condition (4.5) is clearly satisfied by any large enough r and (4.4) reduces to (2.24) so that Theorem 4.2.4 collapses to Theorem 2.3.5 (by taking $K = C$) up to the same small gap between Theorems 4.2.3 and 2.3.4.

Arguing exactly as for Theorem 2.3.7 *a*), it can be shown that the strong pseudomonotonicity of f guarantees that the coercivity condition (4.4) holds provided that the anchorage condition (4.5) is satisfied as well.

Theorem 4.2.5 *Let $K \subseteq \mathbb{R}^n$ satisfy $C(x) \subseteq K$ for any $x \in \mathbb{R}^n$. Suppose $f(x, \cdot)$ is convex for any $x \in \mathbb{R}^n$ and f is μ-pseudomonotone on K with $\mu > 0$. If there exists $r > 0$ such that (4.5) holds for any $x \in \mathbb{R}^n$, then the coercivity condition (4.4) holds for some $r' \geq r$.*

Unlikely the case of Ky Fan inequalities, strong monotonicity does not guarantee the uniqueness of the solution. Indeed, consider QEP(f, C) with $n = 1$, the strongly monotone bifunction $f(x, y) = x(y - x)$ and the variable feasible region $C(x) = [x - 1, x]$. It turns out that fix $C = \mathbb{R}$ while the solution set is $S = (-\infty, 0]$. Notice that the anchorage condition (4.5) is not satisfied.

4.3   Algorithms

Some assumptions on the data of the quasi-equilibrium problem QEP(f, C) are given once and for all in order to provide a unified algorithmic framework throughout all the section. Precisely, the sets $C(x)$ are supposed to be nonempty, closed and convex for any $x \in \mathbb{R}^n$, f to be continuous while $f(x, \cdot)$ to be τ-convex for any $x \in \mathbb{R}^n$ for some $\tau \geq 0$ that does not depend upon the considered point x. Notice that this framework guarantees all the assumptions of the existence Theorem 4.2.3 except for the other assumptions on the map C. Therefore, the existence of at least one solution is generally listed among the assumptions of the convergence theorems.

Fixed-Point and Extragradient Methods

Theorem 4.2.1 shows that QEP(f, C) can be turned into a fixed point problem. Therefore, the same approach developed in Chap. 3 for Ky Fan inequalities can be adapted to the current setting. Indeed, the below algorithm follows in the footsteps of the fixed point algorithm of Sect. 3.1. The only difference lies in the strongly convex optimization problem that is solved at each iteration: both the objective function and the feasible region change according to the current iterate.

Fixed point algorithm for quasi-equilibria

Step 0. Choose $x^0 \in \mathbb{R}^n$, $\alpha > 0$ and set $k = 0$.
Step 1. Compute $x^{k+1} = \arg\min\{f(x^k, y) + \alpha\|y - x^k\|^2/2 \ : \ y \in C(x^k)\}$.
Step 2. If $x^{k+1} = x^k$ then stop.
Step 3. $k = k + 1$ and go to Step 1.

If the algorithm stops at Step 2, then Theorem 4.2.1 guarantees that x^k is a solution of QEP(f, C). Otherwise, suitable assumptions both on f and C are needed to guarantee that QEP(f, C) has a unique solution and the infinite sequence of iterates $\{x^k\}$ converges to this solution with linear rate of convergence.

Since the feasible region of the optimization program at Step 1 changes at each iteration, some Lipschitz control on the behaviour of its unique minimum point is required. More precisely, the functions $y_\alpha(x, \cdot)$, where

$$y_\alpha(x, z) = \arg\min\left\{f(x, y) + \alpha\|y - x\|^2/2 \ : \ y \in C(z)\right\}, \tag{4.6}$$

must satisfy the Lipschitz condition

$$\|y_\alpha(x, z) - y_\alpha(x, z')\| \leq \Lambda(\alpha)\|z - z'\| \tag{4.7}$$

for any $z, z' \in \mathbb{R}^n$ and any $x \in \mathbb{R}^n$ for some $\Lambda(\alpha) \geq 0$ that does not depend upon the considered point x.

Whenever $C(x) = C$ holds for any $x \in \mathbb{R}^n$ for some closed convex set $C \subseteq \mathbb{R}^n$, then (4.7) trivially holds with $\Lambda(\alpha) = 0$ for any $\alpha \in \mathbb{R}$ since the feasible region does not actually change and thus minima don't as well.

If the multivalued map C describes a given set $K \subseteq \mathbb{R}^n$ that moves throughout the space while expanding or retracting in Lipschitz ways, the Lipschitz behaviour of minima (4.7) holds as well.

Theorem 4.3.1 *Let*

$$C(x) = s(x)K + t(x) \tag{4.8}$$

where $K \subset \mathbb{R}^n$ is convex and closed, $t : \mathbb{R}^n \to \mathbb{R}^n$ is Lipschitz continuous with constant L_t, $s : \mathbb{R}^n \to \mathbb{R}$ is Lipschitz continuous with constant L_s and $s(x) > 0$ holds for any $x \in \mathbb{R}^n$. Suppose that $f(x, \cdot)$ is continuously differentiable on $\mathbb{R}^n$ and the map $\nabla_y f(x, \cdot)$ is Lipschitz continuous with constant L for any $x \in \mathbb{R}^n$.

If s is constant, then (4.7) holds with

$$\Lambda(\alpha) = \frac{L_t(|\alpha| + L)}{\alpha + \tau}$$

for any $\alpha > -\tau$.

If K is bounded, then (4.7) holds with

$$\Lambda(\alpha) = \frac{(RL_s + L_t)(|\alpha| + L)}{\alpha + \tau}.$$

for any $\alpha > -\tau$, where $R > 0$ satisfies $K \subseteq \bar{B}(0, R)$.

Proof Consider the $(\alpha + \tau)$-convex function $p(y) = f(x, y) + \alpha \|y - x\|^2/2$ for a given $x \in \mathbb{R}^n$. Then, the first-order optimality conditions for $y_\alpha(x, z)$ and $y_\alpha(x, z')$ guarantee

$$\langle \nabla p(y_\alpha(x, z)), y - y_\alpha(x, z) \rangle \geq 0$$

$$\langle \nabla p(y_\alpha(x, z')), y' - y_\alpha(x, z') \rangle \geq 0$$

to hold for any $y \in C(z)$ and $y' \in C(z')$. Since $s(z)[y_\alpha(x, z') - t(z')]/s(z') + t(z) \in C(z)$ and $s(z')[y_\alpha(x, z) - t(z)]/s(z) + t(z') \in C(z')$, summing the above inequalities with these choices for y and y' gives

$$\langle \nabla p(y_\alpha(x, z)) - \nabla p(y_\alpha(x, z')), s(z)[y_\alpha(x, z') - t(z')] + s(z')[t(z) - y_\alpha(x, z)] \rangle \geq 0.$$

The above inequality and the $(\alpha + \tau)$-monotonicity of ∇p (see Corollary 2.1.12) provide the first two inequalities of the following chain

$$(\alpha + \tau)s(z')\|y_\alpha(x, z) - y_\alpha(x, z')\|^2$$
$$\leq \langle \nabla p(y_\alpha(x, z)) - \nabla p(y_\alpha(x, z')), s(z')[y_\alpha(x, z) - y_\alpha(x, z')] \rangle$$
$$\leq \langle \nabla p(y_\alpha(x, z)) - \nabla p(y_\alpha(x, z')), [s(z) - s(z')][y_\alpha(x, z') - t(z')]$$
$$\quad + s(z')[t(z) - t(z')] \rangle$$
$$\leq \|\nabla p(y_\alpha(x, z)) - \nabla p(y_\alpha(x, z'))\|[|s(z) - s(z')|\|y_\alpha(x, z') - t(z')\|$$
$$\quad + s(z')\|t(z) - t(z')\|]$$
$$\leq (L + |\alpha|)[L_s\|y_\alpha(x, z') - t(z')\| + s(z')L_t]\|y_\alpha(x, z) - y_\alpha(x, z')\|\|z - z'\|,$$

while the third follows from the Cauchy–Schwarz inequality and the last holds since ∇p is Lipschitz continuous with constant $(L + |\alpha|)$. Therefore, the inequality

$$(\alpha + \tau)s(z')\|y_\alpha(x, z) - y_\alpha(x, z')\| \leq (L + |\alpha|)[L_s\|y_\alpha(x, z') - t(z')\| + s(z')L_t]\|z - z'\|$$

holds as well. If s is constant, then $L_s = 0$ and the thesis readily follows. Otherwise, if K is bounded, the thesis follows simply dividing by $s(z')$ and taking into account that $\|y_\alpha(x, z') - t(z')\| \leq Rs(z')$ holds since $y_\alpha(x, z') - t(z') \in s(z')K$.   □

If minima behave in a Lipschitz way according to (4.7) and QEP(f, C) has at least one solution, the same arguments of Sect. 3.1 can be exploited to achieve the convergence of the fixed point algorithm.

Theorem 4.3.2 *Le $\bar{x}$ be a solution of QEP(f, C). Suppose f is μ-monotone, satisfies the inequality (3.1) for any $x, y, z \in \mathbb{R}^n$ for some $T_1, T_2 > 0$ and the Lipschitz behaviour of minima (4.7) holds for some $\alpha \geq 2T_2$. If the fixed point algorithm produces an infinite sequence $\{x^k\}$, then $\bar{x}$ is the unique solution of QEP(f, C) and the whole sequence converges to $\bar{x}$ with the linear rate of convergence*

$$\Lambda(\alpha) + \sqrt{\frac{\alpha + 2T_1 - 2\mu}{\alpha + 2\tau}}, \tag{4.9}$$

provided that this rate is smaller than 1.

Proof The Lipschitz behaviour of minima (4.7) guarantees

$$\begin{aligned}
\|x^{k+1} - \bar{x}\| &= \|y_\alpha(x^k, x^k) - y_\alpha(\bar{x}, \bar{x})\| \\
&\leq \|y_\alpha(x^k, x^k) - y_\alpha(x^k, \bar{x})\| + \|y_\alpha(x^k, \bar{x}) - y_\alpha(\bar{x}, \bar{x})\| \tag{4.10} \\
&\leq \Lambda(\alpha)\|x^k - \bar{x}\| + \|y_\alpha(x^k, \bar{x}) - y_\alpha(\bar{x}, \bar{x})\|.
\end{aligned}$$

Since both $y_\alpha(x^k, \bar{x})$ and $y_\alpha(\bar{x}, \bar{x})$ belong to the set $C(\bar{x})$, the same argument of the proof of Theorem 3.1.1 provides the inequality

$$\|y_\alpha(x^k, \bar{x}) - y_\alpha(\bar{x}, \bar{x})\| \leq \sqrt{\frac{\alpha + 2(T_1 - \mu)}{\alpha + 2\tau}}\|x^k - \bar{x}\|.$$

Combining the above inequalities leads

$$\|x^{k+1} - \bar{x}\| \leq \left(\Lambda(\alpha) + \sqrt{\frac{\alpha + 2T_1 - 2\mu}{\alpha + 2\tau}}\right)\|x^k - \bar{x}\|.$$

Since (4.9) is smaller than 1, the sequence $\{x^k\}$ converges to $\bar{x}$ with the linear rate of convergence (4.9) and hence the solution $\bar{x}$ is unique. □

Monotonicity can be replaced by pseudomonotonicity in the same fashion of Sect. 3.1.

Theorem 4.3.3 *Let $\bar{x}$ be a solution of QEP(f, C). Suppose f is μ-pseudomonotone, satisfies the inequality (3.1) for any $x, y, z \in \mathbb{R}^n$ for some $T_1, T_2 > 0$ such that $\tau + 2\mu - 2T_2 > 0$ and the Lipschitz behaviour of minima (4.7) holds for some $\alpha \geq 2T_1$. If the fixed point algorithm produces an infinite sequence $\{x^k\}$, then $\bar{x}$ is* *the unique solution of QEP(f, C) and the whole sequence converges to $\bar{x}$ with the linear rate of convergence*

$$\Lambda(\alpha) + \sqrt{\frac{\alpha}{\alpha + \tau + 2\mu - 2T_2}},$$

provided that this rate is smaller than 1.

Proof Similarly to the proof of Theorem 4.3.2, inequality (4.10) and the same argument of the proof of Theorem 3.1.2 lead to the thesis. $\square$

Just like for Ky Fan inequalities, the assumptions that guarantee the convergence of the fixed point algorithm imply the uniqueness of the solution of QEP(f, C), which is a rather uncommon feature for quasi-equilibria. Extragradient algorithms can be developed to solve problems with multiple solutions while at the same time getting rid of condition (3.1) through projections onto suitable hyperplanes. Anyway, the existence of a solution with particular features is required to get convergence. More precisely, there must exist a closed convex set $K \subseteq \mathbb{R}^n$ satisfying

$(i) C(z) \subseteq K$ for any $z \in K$,

(ii) there exists $\hat{x} \in \bigcap_{z \in K} C(z)$ such that $f(\hat{x}, y) \geq 0$ for any $y \in K$.   (EAS)

If the inclusions (i) hold, condition (ii) implies that $\hat{x}$ solves EP(f, K) and hence it actually solves QEP(f, C) as well. Moreover, this solution must lie in the intersection of all the sets $C(z)$ over K. Therefore, (i) and (ii) can be together referred to as the existence of an anchor solution $\hat{x}$, shortly (EAS). Notice that the nonemptiness of the above intersection is stronger than the anchorage condition (4.5).

The following examples show that the Lipschitz behaviour (4.7) and the existence of an anchor solution (EAS) are independent of each other.

Example 4.3.4 Consider $f(x, y) = \langle Px, y - x \rangle$ where $P \in \mathbb{R}^{n \times n}$ is positive definite and

$$C(x) = \{y \in \mathbb{R}^n : y_i \leq b_i(x), \quad i = 1, \ldots, n\},$$

where $b_i(x) = \sqrt{|x_i|} - \sqrt{|x_i| + 1} + 1$. Clearly, (EAS) holds with $K = (-\infty, 1]^n$ and $\hat{x} = 0$. On the other hand, minima do not behave in a Lipschitz way. Indeed, consider $n = 2$,

$$P = \begin{pmatrix} 1 & -1 \\ 0 & 1 \end{pmatrix}$$

and $\alpha > 1$. Given any $x \in \text{int } \mathbb{R}_+^2$, there exists a neighborhood U of $\hat{x}$ such that $y_\alpha(x, z) = (b_1(z), b_2(z))$ holds for any $z \in U$. Then, (4.7) does not hold since $y_\alpha(x, \cdot)$ is not Lipschitz continuous on U.

Example 4.3.5 Consider $f(x, y) = \langle Px + Qy, y - x \rangle$ where $Q \in \mathbb{R}^{2 \times 2}$ is positive semidefinite, $P - Q \in \mathbb{R}^{2 \times 2}$ is positive definite and

$$C(x) = \{y \in \mathbb{R}^2 : y_1 \geq 1 - \cos(x_1), \quad y_2 \geq 1 - \cos(x_2)\}.$$

Theorem 4.3.1 guarantees that (4.7) holds since $t(x) = (1 - \cos(x_1), 1 - \cos(x_2))$ provides $C(x) = \mathbb{R}_+^2 + t(x)$. On the other hand, $x = 0$ solves EP(f, K) for any closed convex set $K \subseteq \mathbb{R}^2$ since Q is positive semidefinite and it is the unique solution by Theorem 2.3.7 since f is μ-monotone on $\mathbb{R}^2$ with μ the minimum eigenvalue of $P - Q$. Moreover, any closed convex set K suitable for (EAS) satisfies $\mathbb{R}_+^2 \subseteq K$ since $(2\pi, 2\pi) \in C(x)$ for any $x \in \mathbb{R}^2$. Therefore, the inclusion

$$\bigcap_{z \in K} C(z) \subseteq \bigcap_{z \in \mathbb{R}_+^2} C(z) = [2, +\infty) \times [2, +\infty)$$

shows that the unique solution of EP(f, K) does not lie in the intersection of all $C(z)$ over K. As a consequence, (EAS) does not hold.

Just like for the fixed point algorithm, the hyperplane extragradient algorithm for QEP(f, C) follows in the footsteps of the corresponding algorithm of Sect. 3.1 for Ky Fan inequalities by modifying the feasible region at Step 1 according to the current iterate. Therefore, just as in Sect. 3.1, f is supposed to be continuously differentiable.

Hyperplane extragradient algorithm for quasi-equilibria

Step 0. Choose $x^0 \in K$, $\alpha > -\tau/2$, $\theta, \sigma \in (0, 1)$, let $\beta = \alpha + \tau/2$ and set $k = 0$.

Step 1. Compute $\bar{x}^k = \arg\min\{f(x^k, y) + \alpha\|y - x^k\|^2/2 : y \in C(x^k)\}$.

Step 2. If $\bar{x}^k = x^k$ then stop.

Step 3. Find the smallest $m \in \mathbb{N}$ such that $\theta_k = \theta^m$ and $z^k = (1 - \theta_k)x^k + \theta_k \bar{x}^k$ satisfy

$$f(z^k, x^k) - f(z^k, \bar{x}^k) \geq \beta\sigma\|x^k - \bar{x}^k\|^2.$$

Step 4. Let $d^k = \nabla_y f(z^k, x^k)$, set $\delta_k = f(z^k, x^k)/\|d^k\|^2$ and compute

$$x^{k+1} = P_{C(x^k)}\left(x^k - \delta_k d^k\right).$$

Step 5. $k = k + 1$ and go to Step 1.

The same argument of Sect. 3.1 shows that the line search at Step 3 is well defined and $d^k \neq 0$ holds at Step 4. Moreover, Step 3 amounts to finding an hyperplane

$$H^k = \{x \in \mathbb{R}^n \ : \ f(z^k, x^k) + \langle d^k, x - x^k \rangle = 0\}$$

that separates the current iterate x^k from the subset of the anchor solutions of QEP(f, C), i.e., those solutions satisfying (EAS). Indeed, the convexity of $f(z^k, \cdot)$ implies

$$0 = f(z^k, z^k) \leq (1 - \theta_k) f(z^k, x^k) + \theta_k f(z^k, \bar{x}^k)$$

that coupled with the inequality at Step 3 gives

$$
\begin{aligned}
f(z^k, x^k) + \langle d^k, x^k - x^k \rangle &= f(z^k, x^k) \\
&\geq \theta_k [f(z^k, x^k) - f(z^k, \bar{x}^k)] \\
&\geq \theta_k \beta \sigma \| x^k - \bar{x}^k \|^2 \\
&> 0.
\end{aligned}
$$

On the other hand, both $x^k \in K$ and $\bar{x}^k \in C(x^k) \subset K$ hold so that $z^k \in K$ holds as well. Therefore, $f(\hat{x}, z^k) \geq 0$ holds for any anchor solution $\hat{x}$. If f is pseudomonotone, then the convexity of $f(z^k, \cdot)$ implies

$$f(z^k, x^k) + \langle d^k, \hat{x} - x^k \rangle \leq f(z^k, \hat{x}) \leq 0 \qquad (4.11)$$

and separation actually occurs.

Finally, the stepsize δ_k is chosen in such a way that $x^k - \delta_k d^k$ is the projection of x^k onto H^k and the new iterate x^{k+1} is the projection of such a projection onto $C(x^k)$.

Theorem 4.3.6 *Let (EAS) hold for some compact set K. Suppose f is pseudomonotone and C is both upper and lower semicontinuous. Chosen any $\alpha > -\tau/2$, if the hyperplane extragradient algorithm produces an infinite sequence $\{x^k\}$, then any cluster point of the sequence is a solution of QEP(f, C).*

Proof Let $\hat{x}$ be an anchor solution, i.e., it satisfies (EAS) together with the given K. Then, inequality (4.11) yields

$$\langle d^k, x^k - \hat{x} \rangle \geq f(z^k, x^k) = \delta_k \| d^k \|^2.$$

Since $\hat{x} \in C(x^k)$ and the projection is non-expansive, the following chain of equalities and inequalities

$$\begin{aligned} \|x^{k+1} - \hat{x}\|^2 &= \|P_{C(x^k)}(x^k - \delta_k d^k) - P_{C(x^k)}(\hat{x})\|^2 \\ &\leq \|x^k - \delta_k d^k - \hat{x}\|^2 \\ &= \|x^k - \hat{x}\|^2 - 2\delta_k \langle d^k, x^k - \hat{x} \rangle + \delta_k^2 \|d^k\|^2 \\ &\leq \|x^k - \hat{x}\|^2 - \delta_k^2 \|d^k\|^2 \end{aligned}$$

readily follows. Hence, the sequence $\{\|x^k - \hat{x}\|\}$ is monotone decreasing, bounded by below and thus convergent. As a consequence, the sequence $\{x^k\}$ is bounded.

Since the map $y_\alpha(\cdot)$ is continuous (see Theorem 4.2.2), the same argument of the proof of Theorem 3.1.4 shows that any cluster point of $\{x^k\}$ solves QEP(f, C). $\square$

If a cluster point $\bar{x}$ of the sequence $\{x^k\}$ satisfies (EAS), then the whole sequence converges to $\bar{x}$ since the sequence $\{\|x^k - \hat{x}\|\}$ is convergent with $\hat{x} = \bar{x}$. Notice that it might happen that no cluster point belongs to the intersection required by (EAS), while this is always the case for Ky Fan inequalities as the intersection coincides with the feasible region.

Descent Methods

The quasi-equilibrium problem QEP(f, C) can be solved through any algorithm for the minimization problem (4.3). In order to develop a descent algorithm some assumptions are needed for the regularized gap function φ_α given by (4.2) to be differentiable. Throughout this section the sets $C(x)$ are suppose to be explicitly described by convex constraints, namely,

$$C(x) = \{ y \in \mathbb{R}^n : g_i(x, y) \leq 0 \quad i = 1, \ldots, m \}$$

for some continuously differentiable functions $g_i : \mathbb{R}^n \times \mathbb{R}^n \to \mathbb{R}^n$ such that $g_i(x, \cdot)$ is convex for any $x \in \mathbb{R}^n$. Hence, the set of the fixed points of C is

$$\text{fix } C = \{x \in \mathbb{R}^n : g_i(x, x) \leq 0 \quad i = 1, \ldots, m\}.$$

Notice that the continuity of g_i guarantees that fix C is closed. Furthermore, for the sake of simplicity, the existence of an open set $\Omega \supseteq$ fix C is supposed such that the Slater and linear independence constraint qualifications for the g_i's, namely,

(i) there exists $\hat{y} \in \mathbb{R}^n$ such that $g_i(x, \hat{y}) < 0$ for any $i = 1, \ldots, m$,
(ii) all the vectors $\nabla_y g_i(x, y_\alpha(x))$ such that $g_i(x, y_\alpha(x)) = 0$ are linearly independent,

are satisfied at any $x \in \Omega$, where $y_\alpha(x)$ denotes the unique minimizer of the strongly convex function (4.2) over $C(x)$. Condition *(i)* is equivalent to require that $C(x)$ has a nonempty interior, while condition *(ii)* states the linear independence of the gradients of active constraints at $y_\alpha(x)$.

The above framework guarantees the continuity of the map $y_\alpha(\cdot)$ and the existence of a continuous map $\lambda : \Omega \to \mathbb{R}^m_+$ such that the KKT conditions

$$
\begin{cases}
\nabla_y f(x, y_\alpha(x)) + \alpha(y_\alpha(x) - x) + \displaystyle\sum_{i=1}^m \lambda_i(x)\nabla_y g_i(x, y_\alpha(x)) = 0, \\[2mm]
\lambda_i(x)\, g_i(x, y_\alpha(x)) = 0, \qquad i = 1, \ldots, m
\end{cases}
\tag{4.12}
$$

hold at any $x \in \Omega$ for the optimization problem that provides the regularized gap function φ_α (see Theorems A.2.4 and A.2.5). Moreover, $\lambda(x)$ is the unique vector of multipliers satisfying (4.12) thanks to the linear independence condition *(ii)* (see Theorem A.2.4). This framework guarantees the continuous differentiability of φ_α as well.

Theorem 4.3.7 *The regularized gap function φ_α is continuously differentiable at any point $x \in \mathrm{fix}\, C$ and its gradient is given by*

$$
\nabla\varphi_\alpha(x) = -\nabla_x f(x, y_\alpha(x)) + \alpha(y_\alpha(x) - x) - \sum_{i=1}^m \lambda_i(x)\nabla_x g_i(x, y_\alpha(x)).
\tag{4.13}
$$

Proof Let $x \in \mathrm{fix}\, C$, $d \in \mathbb{R}^n$ and $t > 0$. Since the above constraint qualifications hold at x, the pair $(y_\alpha(x), \lambda(x))$ is a saddle point of the Lagrangian function

$$
L(x, y, \lambda) = f_\alpha(x, y) + \sum_{i=1}^m \lambda_i g_i(x, y),
$$

i.e., the inequalities

$$
L(x, y_\alpha(x), \lambda) \le L(x, y_\alpha(x), \lambda(x)) \le L(x, y, \lambda(x))
$$

hold for any $y \in \mathbb{R}^n$ and any $\lambda \in \mathbb{R}^m_+$. Thanks to the complementary slackness conditions, $L(x, y_\alpha(x), \lambda(x)) = f_\alpha(x, y_\alpha(x)) = -\varphi_\alpha(x)$ holds and the above inequalities can be equivalently turned into

$$
-L(x, y, \lambda(x)) \le \varphi_\alpha(x) \le -L(x, y_\alpha(x), \lambda).
\tag{4.14}
$$

Similarly, replacing x by $x_t = x + td$ provides

$$
-L(x_t, y, \lambda(x_t)) \le \varphi_\alpha(x_t) \le -L(x_t, y_\alpha(x_t), \lambda).
\tag{4.15}
$$

Setting $y = y_\alpha(x_t)$ in the left inequality of (4.14) and $\lambda = \lambda(x)$ in the right inequality of (4.15), the mean value theorem leads to

$$\frac{\varphi_\alpha(x_t) - \varphi_\alpha(x)}{t} \leq [L(x, y_\alpha(x_t), \lambda(x)) - L(x_t, y_\alpha(x_t), \lambda(x))]/t$$

$$= \left\{ f_\alpha(x, y_\alpha(x_t)) - f_\alpha(x_t, y_\alpha(x_t)) \right.$$

$$\left. + \sum_{i=1}^m \lambda_i(x)[g_i(x, y_\alpha(x_t)) - g_i(x_t, y_\alpha(x_t))] \right\}/t$$

$$= -\langle \nabla_x f_\alpha(\xi_t, y_\alpha(x_t)), d \rangle - \sum_{i=1}^m \lambda_i(x)\langle \nabla_x g_i(\eta_t^i, y_\alpha(x_t)), d \rangle$$

$$= \langle -\nabla_x f(\xi_t, y_\alpha(x_t)) + \alpha(y_\alpha(x_t) - \xi_t) - \sum_{i=1}^m \lambda_i(x)\nabla_x g_i(\eta_t^i, y_\alpha(x_t)), d \rangle$$

where ξ_t and η_t^i belong to the line segment between x and x_t for any $i = 1, \dots, m$. The continuity of the maps $y_\alpha(\cdot)$, $\nabla_x f$ and $\nabla_x g_i$ implies

$$\limsup_{t \to 0^+} \frac{\varphi_\alpha(x_t) - \varphi_\alpha(x)}{t} \leq$$

$$\langle -\nabla_x f(x, y_\alpha(x)) + \alpha(y_\alpha(x) - x) - \sum_{i=1}^m \lambda_i(x)\nabla_x g_i(x, y_\alpha(x)), d \rangle.$$

Similarly, setting $\lambda = \lambda(x_t)$ in the right inequality of (4.14) and $y = y_\alpha(x)$ in the left inequality of (4.15), the mean value theorem and the continuity of the maps $\lambda(\cdot)$, $\nabla_x f$ and $\nabla_x g_i$ guarantee the following inequality

$$\liminf_{t \to 0^+} \frac{\varphi_\alpha(x_t) - \varphi_\alpha(x)}{t} \geq$$

$$\langle -\nabla_x f(x, y_\alpha(x)) + \alpha(y_\alpha(x) - x) - \sum_{i=1}^m \lambda_i(x)\nabla_x g_i(x, y_\alpha(x)), d \rangle.$$

The two above inequalities imply that the one-sided directional derivative of φ_α at x in the direction d exists and it is given by

$$\varphi_\alpha'(x; d) = \langle -\nabla_x f(x, y_\alpha(x)) + \alpha(y_\alpha(x) - x) - \sum_{i=1}^m \lambda_i(x)\nabla_x g_i(x, y_\alpha(x)), d \rangle.$$

Finally, the continuity of the maps $y_\alpha(\cdot)$, $\lambda(\cdot)$, $\nabla_x f$ and $\nabla_x g_i$ implies that φ_α is continuously differentiable at x with the gradient given by (4.13). $\square$

A descent direction for φ_α can be obtained under suitable concavity conditions on f by adding a technical assumption on the constraining functions.

Theorem 4.3.8 *Let $x \in$ fix C and $\alpha > -\tau$. Suppose fix C is convex, $y_\alpha(x) \in$ fix C, $f(\cdot, y_\alpha(x))$ is γ-concave for some $\gamma > -\tau$ and the constraining functions g_i satisfy*

$$\langle \nabla_x g_i(x, y_\alpha(x)) + \nabla_y g_i(x, y_\alpha(x)), y_\alpha(x) - x \rangle \geq 0 \quad \text{if } g_i(x, y_\alpha(x)) = 0. \tag{4.16}$$

If x is not a solution of QEP(f, C), then the direction $d = y_\alpha(x) - x$ satisfies

$$\langle \nabla \varphi_\alpha(x), d \rangle \leq -(\gamma + \tau)\|d\|^2/2. \tag{4.17}$$

Proof Theorem 4.2.1 guarantees $d \neq 0$. Hence, the following chain of equalities and inequalities hold:

$$
\begin{aligned}
\langle \nabla \varphi_\alpha(x), d \rangle &= \langle -\nabla_x f(x, y_\alpha(x)) + \alpha d - \sum_{i=1}^m \lambda_i \nabla_x g_i(x, y_\alpha(x)), d \rangle \\
&\leq f(x, y_\alpha(x)) + (\alpha - \gamma/2)\|d\|^2 - \sum_{i=1}^m \lambda_i \langle \nabla_x g_i(x, y_\alpha(x)), d \rangle \\
&\leq f(x, y_\alpha(x)) + (\alpha - \gamma/2)\|d\|^2 + \sum_{i=1}^m \lambda_i \langle \nabla_y g_i(x, y_\alpha(x)), d \rangle \\
&= f(x, y_\alpha(x)) - \gamma\|d\|^2/2 - \langle \nabla_y f(x, y_\alpha(x)), d \rangle \\
&\leq -(\gamma + \tau)\|d\|^2/2.
\end{aligned}
$$

where $\lambda \in \mathbb{R}_+^m$ satisfies the optimality conditions (4.12) together with x. The first inequality comes from γ-concavity (see (3.11)), the second from (4.16) paired with the complementarity slackness conditions in (4.12), the third from τ-convexity, while the last equality is due to the multiplier rule in (4.12).   □

When Ky Fan inequalities are considered, the convex functions g_i do not depend upon x and the technical assumption (4.16) is always satisfied due to the convexity of the constraining functions g_i's. Therefore, Theorem 4.3.8 collapses to the descent property given in Theorem 3.2.2.

The descent direction of Theorem 4.3.8 can be exploited to devise a descent algorithm in the same fashion of Sect. 3.2. Just like the two previous algorithms, the only difference with the corresponding algorithm for Ky Fan Inequalities lies in the dependence of feasible region of the convex optimization problem at Step 1 from the current iterate.

Descent algorithm for quasi-equilibria

Step 0. Choose $\alpha > -\tau$, $\eta \in (0, [\gamma + \tau]/2)$, $\delta \in (0, 1)$, $x^0 \in$ fix C and set $k = 0$.
Step 1. Compute $y_\alpha(x^k) = \arg\min\{f(x^k, y) + \alpha\|y - x^k\|^2/2 \; : \; y \in C(x^k)\}$.
Step 2. If $d^k = y_\alpha(x^k) - x^k = 0$ then stop.
Step 3. Compute the smallest non-negative integer s such that

$$\varphi_\alpha(x^k + \delta^s d^k) - \varphi_\alpha(x^k) \leq -\eta\,\delta^s\,\|d^k\|^2.$$

Step 4. Set $t_k = \delta^s$, $x^{k+1} = x^k + t_k d^k$, $k = k + 1$ and go to Step 1.

If the algorithms stops at Step 2, then the current iterate solves QEP(f, C) by Theorem 4.2.1. Moreover, Theorem 4.3.8 and the choice of η guarantee that the line search procedure at Step 3 is finite. In fact, suppose by contradiction that some iteration k satisfies

$$\varphi_\alpha(x^k + \delta^s d^k) - \varphi_\alpha(x^k) > -\eta\,\delta^s\,\|d^k\|^2$$

for any $s \in \mathbb{N}$. Then, taking the limit as $s \to +\infty$ leads to

$$\langle \nabla\varphi_\alpha(x^k), d^k \rangle \geq -\eta\,\|d^k\|^2 > -(\gamma + \tau)\,\|d^k\|^2/2$$

that contradicts (4.17) since x^k does not satisfy the stopping criterion of Step 2 and hence does not solve QEP(f, C).

Even though the current iterate x^k is a fixed point of C there is no guarantee that the new one is a fixed point as well unless some additional assumptions are made. In order to guarantee convergence to a solution of QEP(f, C), C is supposed to map $\mathbb{R}^n$ into fix C (so that $y_\alpha(x^k) \in$ fix C) and this latter set to be convex, which holds true if the g_i's are jointly quasiconvex in both variables. Under these assumptions all the iterates are fixed points of C.

Theorem 4.3.9 *Suppose* fix C *is convex and* $C(x) \subseteq$ fix C *holds for any* $x \in$ fix C, $f(\cdot, y)$ *is* γ-*concave for any* $y \in$ fix C *for some* $\gamma > -\tau$ *and the constraining functions* g_i *satisfy* (4.16) *for any* $x \in$ fix C. *If the descent algorithm produces an infinite sequence* $\{x^k\}$, *then any of its cluster points solves* QEP(f, C).

Proof Let $\bar{x}$ be any cluster point of the sequence $\{x^k\}$ and let $x^{k_\ell} \to \bar{x}$ for some $k_\ell \uparrow +\infty$ as $\ell \uparrow +\infty$. Since fix C is a closed set, then $\bar{x} \in$ fix C. The continuity of $y_\alpha(\cdot)$ guarantees $d^{k_\ell} = y_\alpha(x^{k_\ell}) - x^{k_\ell} \to \bar{d} = y_\alpha(\bar{x}) - \bar{x}$. Hence, $\bar{d} = 0$ guarantees that $\bar{x}$ solves QEP(f, C) by Theorem 4.2.1. By contradiction, suppose $\bar{d} \neq 0$. The same argument of the proof of Theorem 3.2.1 with φ_α in place of ψ and $\sigma_{k_\ell} = \|d^{k_\ell}\|^2$ leads to

$$\langle \nabla\varphi_\alpha(\bar{x}), \bar{d} \rangle \geq -\eta\,\|\bar{d}\|^2 > -(\gamma + \tau)\,\|\bar{d}\|^2/2.$$

Thus, Theorem 4.3.8 implies that $\bar{x}$ solves QEP(f, C), otherwise

$$\langle \nabla \varphi_\alpha(\bar{x}), \bar{d} \rangle \leq -(\gamma + \tau) \|\bar{d}\|^2 / 2$$

should hold. $\square$

The technical condition (4.16) on the gradients of the constraining functions is a key tool to control the descent through the bound (4.17) and to achieve the convergence. Fortunately, there are a few classes of constraints that actually meet it.

Theorem 4.3.10 *Let c_i be convex and continuously differentiable, $F_i : \mathbb{R}^n \to \mathbb{R}^n$ be monotone and differentiable, $R_i \in \mathbb{R}^{n \times n}$ be positive semidefinite, $Q_i \in \mathbb{R}^{n \times p}$, $a, r_i \in \mathbb{R}^n$ and $\beta_i \in \mathbb{R}$, $\alpha, \alpha_i \leq 1$, $v_i \leq 0$. Then, any of the following bifunctions*

a) $g_i(x, y) = c_i(y) - c_i(x)$
b) $g_i(x, y) = c_i(Q_i [y - (\alpha x + a)])$
c) $g_i(x, y) = \langle F_i(x) + R_i y + r_i, y - x \rangle + v_i$
d) $g_i(x, y) = y_i - \alpha_i x_i - \beta_i$
e) $g_i(x, y) = \alpha_i x_i + \beta_i - y_i$

satisfies the condition (4.16) at any $x \in$ fix C.

Proof Given any $x \in$ fix C, suppose $y = y_\alpha(x)$ satisfies $g_i(x, y) = 0$.
a) The convexity of c_i implies that ∇c_i is monotone, that guarantees

$$\langle \nabla_x g_i(x, y) + \nabla_y g_i(x, y), y - x \rangle = \langle \nabla c_i(y) - \nabla c_i(x), y - x \rangle \geq 0.$$

b) The convexity of c_i guarantees that $g_i(x, \cdot)$ is convex so that (4.16) can be obtained in the following way

$$
\begin{aligned}
\langle \nabla_x g_i(x, y) + \nabla_y g_i(x, y), y - x \rangle &= (1 - \alpha) \langle Q_i \nabla c_i(Q_i [y - (\alpha x + a)]), y - x \rangle \\
&= (1 - \alpha) \langle \nabla_y g_i(x, y), y - x \rangle \\
&\geq (1 - \alpha) [g_i(x, y) - g_i(x, x)] \\
&= -(1 - \alpha) g_i(x, x) \\
&\geq 0,
\end{aligned}
$$

where the last inequality holds since $x \in$ fix C requires $g_i(x, x) \leq 0$.
c) The monotonicity of F_i implies that $\nabla F_i(x)$ is positive semidefinite so that

$$\langle \nabla_x g_i(x, y) + \nabla_y g_i(x, y), y - x \rangle = \langle y - x, [\nabla F_i(x) + R_i](y - x) \rangle \geq 0.$$

holds since R_i is positive semidefinite as well.

d) In this case $g_i(x, y) = 0$ and $x \in \text{fix } C$ read $x_i \leq \alpha_i x_i + \beta_i = y_i$. Therefore, $\alpha_i \leq 1$ guarantees

$$\langle \nabla_x g_i(x, y) + \nabla_y g_i(x, y), y - x \rangle = (1 - \alpha_i)(y_i - x_i) \geq 0.$$

e) Analogous to *d)*.   □

Beyond box constraints, more general linear constraints with variable right-hand side, i.e., $g_i(x, y) = \langle d, y \rangle - c_i(x)$, satisfy (4.16) at any $x \in \text{fix } C$ provided that c_i is a convex function such that $C(x)$ includes the sublevel set constraint $c_i(y) \leq c_i(x)$ as well. Indeed, the following chain of equalities and inequalities hold

$$\begin{aligned}
\langle \nabla_x g_i(x, y) + \nabla_y g_i(x, y), y - x \rangle &= \langle d - \nabla c_i(x), y - x \rangle \\
&= -\langle \nabla c_i(x), y - x \rangle + (c_i(x) - \langle d, x \rangle) \\
&\geq -\langle \nabla c_i(x), y - x \rangle \\
&\geq c_i(x) - c_i(y) \\
&\geq 0.
\end{aligned}$$

Shared constraints are a frequent feature in noncooperative games. A linear equality shared constraint is given by

$$\langle a_1, x_1 \rangle + \cdots + \langle a_\ell, x_\ell \rangle = b \tag{4.18}$$

for some $a_i \in \mathbb{R}^{n_i}$, $b \in \mathbb{R}$. In the reformulation of a generalized Nash game as $\text{QEP}(f, C)$ through the Nikaido–Isoda bifunction (see Sect. 4.1) it can be described by the 2ℓ bifunctions

$$g_i(x, y) = \langle a_i, y_i \rangle + \sum_{j \neq i} \langle a_j, x_j \rangle - b, \qquad i = 1, \ldots, \ell,$$

and $-g_i(x, y)$. All the above constraints satisfy condition (4.16) at any $x \in \text{fix } C$. In fact, $x \in \text{fix } C$ guarantees that (4.18) holds so that any $y \in C(x)$ satisfies

$$\langle a_i, y_i \rangle - \langle a_i, x_i \rangle = b - \sum_{j \neq i} \langle a_j, x_j \rangle - \langle a_i, x_i \rangle = b - \sum_{j=i}^{\ell} \langle a_i, x_i \rangle = 0.$$

Therefore, the equalities

$$\langle \nabla_x g_i(x, y) + \nabla_y g_i(x, y), y - x \rangle = \sum_{j=1}^{\ell} (\langle a_j, y_j \rangle - \langle a_j, x_j \rangle) = 0$$

hold for any $y \in C(x)$ as $g_i(x, y) = 0$ for all i's.

Moreover, the condition $y_\alpha(x) \in \text{fix } C$ is guaranteed as well if there are no other constraints since any $y \in C(x)$ satisfies

$$\sum_{j=1}^{\ell} \langle a_j, y_j \rangle = \sum_{j=1}^{\ell} \langle a_j, x_j \rangle = b.$$

If all the constraints are of kind b) of Theorem 4.3.10, then $C(x) = K + t(x)$ is the moving set provided by the convex set $K = \bigcap_i \{ y \in \mathbb{R}^n \ : \ c_i(Q_i y) \le 0 \}$ together with $t(x) = \alpha x + a$.

If all the constraints g_i fall within the kinds a) and c), then fix $C = \mathbb{R}^n$. This is no longer true if some of them are of the kind b), d) or e). In the case b) the set of the fixed points turns out to be

$$\text{fix } C = \begin{cases} (K+a)/(1-\alpha) & \text{if } \alpha < 1, \\ \varnothing & \text{if } \alpha = 1 \text{ and } -a \notin K, \\ \mathbb{R}^n & \text{if } \alpha = 1 \text{ and } -a \in K. \end{cases}$$

In this case $C(x) \subseteq \text{fix } C$ holds and therefore also $y_\alpha(x) \in \text{fix } C$ hold for any $x \in \text{fix } C$ whenever fix C is nonempty. Indeed, if $\alpha < 1$, then

$$\begin{aligned} (1-\alpha)C(x) &= (1-\alpha)K + \alpha(1-\alpha)x + (1-\alpha)a \\ &\subseteq (1-\alpha)K + \alpha(K+a) + (1-\alpha)a \\ &= K + a. \end{aligned}$$

If all the variables y_i's are bounded by above through box constraints of kind d) with $\alpha_i < 1$ and there are no other constraints, then fix $C = \prod_i (-\infty, \beta_i/(1-\alpha_i)]$ is a box unbounded by below. Similarly, fix C is a box unbounded by above if only constraints of kind e) exist while it is a bounded box if both kinds of constraints describe the feasible region. In all these situations $x \in \text{fix } C$ guarantees also $y_\alpha(x) \in \text{fix } C$ whenever $\alpha_i > 0$ holds for all i's.

Theorem 4.3.9 does not guarantee the existence of a cluster point of the sequence generated by the descent algorithm. Anyway, suitable Lipschitz and monotonicity assumptions guarantee the sublevels sets of the regularized gap function φ_α are bounded so that cluster points do exist in this framework.

Theorem 4.3.11 *Let $\bar{x}$ be a solution of QEP(f, C). Suppose f is μ-monotone with $\mu > 0$, there exist $L_1, L_2 > 0$ such that*

$$\| \nabla_y f(x, z) \| \le L_1 \quad \text{and} \quad \| \nabla_y f(x, z) - \nabla_y f(\bar{x}, z) \| \le L_2 \| x - \bar{x} \| \qquad (4.19)$$

hold for any $x \in \text{fix } C$ and $z \in C(x)$, and the Lipschitz behaviour of minima (4.7) holds for some $\alpha > -\tau$. Then, the sublevel sets $\{ x \in \text{fix } C \ : \ \varphi_\alpha(x) \le c \}$ are bounded for any $c > 0$.

Proof Given any $x \in$ fix C, the first-order optimality conditions for $y_\alpha(x)$ imply

$$\langle \nabla_y f(x, y_\alpha(x)), x - y_\alpha(x) \rangle \geq \alpha \| y_\alpha(x) - x \|^2. \tag{4.20}$$

Then, the following chain of equalities and inequalities holds:

$$
\begin{aligned}
0 &\leq \langle \nabla_y f(x, y_\alpha(x)), x - y_\alpha(x) \rangle \\
&= \langle \nabla_y f(x, y_\alpha(x)) - \nabla_y f(\bar{x}, y_\alpha(x)) + \nabla_y f(\bar{x}, y_\alpha(x)), x - y_\alpha(x) \rangle \\
&\leq \| \nabla_y f(x, y_\alpha(x)) - \nabla_y f(\bar{x}, y_\alpha(x)) \| \, \| x - y_\alpha(x) \| + f(\bar{x}, x) - f(\bar{x}, y_\alpha(x)) \\
&\leq L_2 \| x - \bar{x} \| \, \| x - y_\alpha(x) \| + f(\bar{x}, x) + f(\bar{x}, y_\alpha(x, \bar{x})) - f(\bar{x}, y_\alpha(x)) \\
&\leq L_2 \| x - \bar{x} \| \, \| x - y_\alpha(x) \| - f(x, \bar{x}) - \mu \| x - \bar{x} \|^2 \\
&\quad + \langle \nabla_y f(\bar{x}, y_\alpha(x, \bar{x})), y_\alpha(x, \bar{x}) - y_\alpha(x) \rangle \\
&\leq L_2 \| x - \bar{x} \| \, \| x - y_\alpha(x) \| - f(x, \bar{x}) - \mu \| x - \bar{x} \|^2 \\
&\quad + \| \nabla_y f(\bar{x}, y_\alpha(x, \bar{x})) \| \, \| y_\alpha(x, \bar{x}) - y_\alpha(x) \| \\
&\leq L_2 \| x - \bar{x} \| \, \| x - y_\alpha(x) \| + \langle \nabla_y f(x, x), x - \bar{x} \rangle - \mu \| x - \bar{x} \|^2 + L_1 \Lambda(\alpha) \| x - \bar{x} \| \\
&\leq L_2 \| x - \bar{x} \| \, \| x - y_\alpha(x) \| + L_1 \| x - \bar{x} \| - \mu \| x - \bar{x} \|^2 + L_1 \Lambda(\alpha) \| x - \bar{x} \|.
\end{aligned}
$$

The second inequality follows from the Cauchy–Schwarz inequality coupled with the convexity of $f(\bar{x}, \cdot)$, the third from (4.19) taking into account also that $y_\alpha(x, \bar{x}) \in C(\bar{x})$ and $\bar{x}$ solves QEP(f, C). The fourth inequality follows from the μ-monotonicity and the convexity of $f(\bar{x}, \cdot)$, the fifth again from the Cauchy–Schwarz inequality while the last two from the convexity of $f(x, \cdot)$, (4.7) and (4.19). Therefore, the inequality

$$\| x - \bar{x} \| \leq [L_2 \| x - y_\alpha(x) \| + L_1 + L_1 \Lambda(\alpha)] / \mu$$

holds as well. Moreover, Lemma 2.5.1 with $g = f(x, \cdot)$, $\xi = x$, $C = C(x)$ and $y^\star = y_\alpha(x)$ implies

$$\varphi_\alpha(x) = -f(x, y_\alpha(x)) - \alpha \| y_\alpha(x) - x \|^2 / 2 \geq (\alpha + \tau) \| y_\alpha(x) - x \|^2 / 2.$$

Therefore, the error bound

$$\| x - \bar{x} \| \leq [L_2 \sqrt{2 \varphi_\alpha(x)/(\alpha + \tau)} + L_1 (1 + \Lambda(\alpha))] / \mu$$

implies the thesis. $\qquad\square$

4.4   Notes and References

Unlike Ky Fan inequalities, generalized Nash equilibrium problems (GNEPs) and quasi-variational inequalities (QVIs), the quasi-equilibrium format of this chapter (shortly, the QEP format) did not yet receive much attention.

QVIs have been introduced in [21, 22] to investigate impulse control problems. Later, they have been exploited to model finite and infinite-dimensional equilibrium problems (see, for instance, [17, 51, 96]). Several algorithmic approaches have been devised for QVIs: fixed points and projections methods [51, 101, 105, 121], penalization of coupling constraints [62, 109, 116], solution of suitable KKT systems [64] and Newton type methods [107, 108].

GNEPs have been introduced in [13, 56] and exploited in many fields (see, for instance, the survey paper [62]). Mainly in the last decade, different kinds of algorithms for GNEPs have been developed: projection methods [134], penalty and barrier methods [86], locally convergent Newton [63, 74] and globally convergent interior-point algorithms [61].

The reformulation of a quasi-equilibrium problem as an optimization problem through a suitable gap function (Theorem 4.2.1) has been considered in [32] relying on a general auxiliary term. Theorem 4.2.2 on the continuity of the regularized gap function is the analogous of Theorem 2.2.6 for Ky Fan inequalities and its proof follows the footsteps of the Berge's Maximum Theorem [23].

There are only few results about the existence of solutions of QEPs in finite dimensional spaces. The technique exploited in Theorem 4.2.3 (see [49]) is similar to the one originally introduced in [53] and it relies on the continuous selection Theorem A.3.10 (see [94]). The main difference between the two results is that Theorem 4.2.3 requires the upper semicontinuity of f separately in the two variables and not globally: in this way Theorem 4.2.3 essentially collapses to Theorem 2.3.4 if the quasi-equilibrium problem turns out to be the Ky Fan inequality. By combining fixed point techniques with the stability analysis of perturbed equilibrium problems, an existence result has been achieved in [16] under quasimonotonicity of f and a weak continuity condition. Theorem 4.2.4 is an extension of Theorem 4.2.3 to the case of unbounded feasible regions and it is in the vein of the results in [53].

The fixed point algorithm of Sect. 4.3 seems to be the most natural approach for solving QEPs, but surprisingly it has not been investigated up to very recently (see [35]). The Lipschitz behaviour of minima (4.7) has been introduced in [32] to get an error bound for the regularized gap function. The hyperplane extragradient algorithm has been devised in [124] and further developed in [126]. It is based on the existence of an anchor solution (EAS), an assumption that has been originally introduced in [134] for GNEPs without naming it.

The descent approach of Sect. 4.3 has been proposed in [32]. The technical assumption (4.16) can be actually considered as a monotonicity type condition (see, for instance, [36]) and the analysis of the classes of constraints that satisfy it is taken from [32]. Moreover, constraint qualifications stronger than needed are considered in this chapter for the sake of simplicity. In fact, if the linear independence of the gradients of active constraints is dropped, the regularized gap function may not be continuously differentiable and the generalized Clarke derivative has to be exploited in order to achieve the convergence of the algorithm.

Appendix A
Mathematical Background

A.1 Topological Concepts

The *inner product* of the vectors $x = (x_1, \ldots, x_n)$ and $y = (y_1, \ldots, y_n)$ is

$$\langle x, y \rangle = \sum_{i=1}^{n} x_i y_i.$$

The inner product naturally induces the *(Euclidean) norm* that measures the length of a vector. The norm of $x \in \mathbb{R}^n$ is defined by

$$\|x\| = \sqrt{\langle x, x \rangle} = \sqrt{\sum_{i=1}^{n} x_i^2}$$

and it represents the distance of x from the origin. The main properties of the norm, which can be deduced from the properties of the inner product, are the following:

- $\|x\| \geq 0$ for all $x \in \mathbb{R}^n$ and $\|x\| = 0$ if and only if $x = 0$,
- $\|\alpha x\| = |\alpha| \cdot \|x\|$ for all $x \in \mathbb{R}^n$ and $\alpha \in \mathbb{R}$,
- $\|x + y\| \leq \|x\| + \|y\|$ for all $x, y \in \mathbb{R}^n$ (triangle inequality)

The *distance* between two points $x, y \in \mathbb{R}^n$ is given by

$$\|x - y\| = \sqrt{\sum_{i=1}^{n} (x_i - y_i)^2}.$$

© Springer Nature Switzerland AG 2019
G. Bigi et al., *Nonlinear Programming Techniques for Equilibria*, EURO Advanced
Tutorials on Operational Research, https://doi.org/10.1007/978-3-030-00205-3

An important property of the inner product follows directly from the law of cosines:

$$\cos \theta = \frac{\langle x, y \rangle}{\|x\| \cdot \|y\|},$$

where $\theta \in [0, \pi]$ is the angle between two non-zero vectors x and y. As a consequence, the inner product of two orthogonal vectors is zero and, since $|\cos \theta| \leq 1$, the following well-known *Cauchy-Schwarz inequality* holds

$$|\langle x, y \rangle| \leq \|x\| \cdot \|y\|$$

for every $x, y \in \mathbb{R}^n$. The *ball* of radius $r > 0$ and centered at $\bar{x} \in \mathbb{R}^n$ is the set

$$B(\bar{x}, r) = \{x \in \mathbb{R}^n : \|x - \bar{x}\| < r\},$$

while the *closed ball* is

$$\bar{B}(\bar{x}, r) = \{x \in \mathbb{R}^n : \|x - \bar{x}\| \leq r\}.$$

A *neighborhood* of $x \in \mathbb{R}^n$ is any subset $U \subseteq \mathbb{R}^n$ that includes a suitable ball $B(x, r)$. The *diameter* of a set X is the quantity

$$\text{diam}(X) = \sup\{\|x - y\| : x, y \in X\}$$

and the set X is bounded if and only if there exists $r > 0$ such that $X \subseteq B(0, r)$ or equivalently $\text{diam}(X) \leq 2r$. The *(Minkowski) sum* of two subsets X and Y of $\mathbb{R}^n$ is

$$X + Y = \{x + y : x \in X, \ y \in Y\}.$$

Similarly, the multiplication of a number $a \in \mathbb{R}$ and a set $X \subseteq \mathbb{R}^n$ provides the set $aX = \{ax : x \in X\}$. If X is a convex set, then $aX + bX = (a+b)X$. The equivalence does not hold if X is not convex: indeed if $X = \{-1, 1\}$ then $X + X = \{-2, 0, 2\}$ and $2X = \{-2, 2\}$.

A sequence $\{x^k\} \subset \mathbb{R}^n$ *converges* to the limit $x \in \mathbb{R}^n$, and it is denoted by either $x^k \to x$ or $\lim_{k \to +\infty} x^k = x$, if for any neighborhood U of x there exists $\bar{k} \in \mathbb{N}$ such that $x^k \in U$ for all $k > \bar{k}$. Any convergent sequence is clearly bounded. A sequence $\{x^k\}$ is convergent if and only if it is a *Cauchy sequence*, i.e., given any $\varepsilon > 0$ there exists $\bar{k} \in \mathbb{N}$ such that $\|x^i - x^j\| < \varepsilon$ for any pair of natural numbers $i, j > \bar{k}$.

A *subsequence* of $\{x^k\}$ is any sequence of the form $\{x^{k_\ell}\}$ where $\{k_\ell\}$ is a strictly increasing sequence of positive integers. Every subsequence of a convergent sequence is convergent to the same limit. A point $x \in \mathbb{R}^n$ is a *cluster point* of the sequence $\{x^k\}$ if, given any neighborhood U of x, there exist infinitely many $k \in \mathbb{N}$ such that $x^k \in U$. This is equivalent to state that x is a limit of some subsequence $\{x^{k_\ell}\}$. When the sequences are of real numbers, the infimum and supremum of its cluster points are called *limit inferior* and *limit superior* of the sequence, respectively. More formally:

– the limit inferior of $\{x_k\} \subset \mathbb{R}$ is

$$\liminf_{k \to +\infty} x_k = \inf\{L \in [-\infty, +\infty] : \text{there is } \{x_{k_\ell}\} \text{ such that } x_{k_\ell} \to L\};$$

– the limit superior of $\{x_k\} \subset \mathbb{R}$ is

$$\limsup_{k \to +\infty} x_k = \sup\{L \in [-\infty, +\infty] : \text{there is } \{x_{k_\ell}\} \text{ such that } x_{k_\ell} \to L\}.$$

Notice that the terms of a sequence of real numbers are written x_k rather than x^k. The sequence $\{x_k\} \subset \mathbb{R}$ converges to L if and only if $\liminf_{k \to +\infty} x_k = \limsup_{k \to +\infty} x_k = L$.

Some topological properties of the subsets of $\mathbb{R}^n$ are listed below. A set $X \subseteq \mathbb{R}^n$ is said to be *open* if for any $x \in X$ there exists a neighborhood U of x such that $U \subseteq X$ and it is *closed* if its complement X^c is an open set. An alternative characterization of closed sets is available via sequences: the set X is closed if and only if the limit of any convergent sequence of elements of X belongs to X as well. By definition the sets $\emptyset$ and $\mathbb{R}^n$ are both open and closed and it is possible to show no other set is open and closed at the same time. Moreover, the intersection of an arbitrary family of closed sets is closed. The *interior* of X, denoted by int X, is the largest open set contained in X, that is

$$\text{int } X = \{x \in X : \exists U \subseteq X \text{ such that } U \text{ is a neighborhood of } x\};$$

the *closure* of X, denoted by cl X, is the smallest closed set containing X, that is

$$\text{cl } X = \{x \in \mathbb{R}^n : \exists \{x^k\} \subseteq X \text{ such that } x^k \to x\}.$$

The *boundary* of X is the set of points in the closure of X that do not belong to the interior of X, or equivalently it is the intersection of the closure of X with the closure of its complement, that is

$$\partial X = \text{cl } X \setminus \text{int } X = \text{cl } X \cap \text{cl } X^c.$$

In many situations the topological properties of subsets $X \subseteq Y$ are required to be considered relatively to a given subset Y of $\mathbb{R}^n$ and not with respect to $\mathbb{R}^n$. X is said to be *relatively open* in Y if it is the intersection of Y with an open subset of $\mathbb{R}^n$, *relatively closed* in Y if it is the intersection of Y with a closed subset of $\mathbb{R}^n$. Whenever Y is open, X is relatively open in Y if and only if X is open. Whenever Y is closed, X is relatively closed in Y if and only if X is closed. Clearly, $\emptyset$ and Y are both relatively open and closed in Y. They are the unique sets with both properties hold if in addition Y is convex.

The definitions of interior, closure and boundary of X in Y read as follows:

- the interior of X in Y is the largest relatively open set in Y contained in X, that is

$$\mathrm{int}_Y\, X = \{x \in X :\ \exists U \text{ neighborhood of } x \text{ such that } U \cap Y \subseteq X\};$$

- the closure of X in Y is the smallest relatively closed set in Y containing X, that is

$$\mathrm{cl}_Y\, X = \{x \in Y :\ \exists \{x^k\} \subseteq X \text{ such that } x^k \to x\};$$

- the boundary of X in Y is $\partial_Y X = \mathrm{cl}_Y\, X \setminus \mathrm{int}_Y\, X = \mathrm{cl}_Y\, X \cap (Y \setminus \mathrm{cl}_Y\, X)$.

Compact sets play a fundamental role in the existence of solution of an optimization problem and a few equivalent definitions may be given.

A set $X \subseteq \mathbb{R}^n$ is called *compact* if one of the following equivalent conditions is satisfied:

- given any family $\{\Omega_i\}_{i \in I}$ of open sets such that $X \subseteq \bigcup_{i \in I} \Omega_i$, there exists a finite subset $J \subseteq I$ such that $X \subseteq \bigcup_{i \in J} \Omega_i$;
- any collection $\{C_i\}_{i \in I}$ of closed subsets of X with the *finite intersection property*, i.e., $\bigcap_{i \in J} C_i \neq \emptyset$ for any finite $J \subseteq I$, has a nonempty intersection;
- X is closed and bounded;
- any sequence in X has a convergent subsequence whose limit belongs to X.

Since intersection of closed sets is closed, the intersection of a closed set and a compact set is still compact.

Given $\{x^i\}_{i \in I} \subseteq \mathbb{R}^n$ for a finite index set I, the set of all the convex combinations of all the points x^i, that is

$$S = \mathrm{conv}\left(\{x^i\}_{i \in I}\right) = \left\{ x = \sum_{i \in I} \lambda_i x^i : \lambda_i \geq 0 \ \text{ and } \ \sum_{i \in I} \lambda_i = 1 \right\},$$

is called the convex hull of $\{x^i\}_{i \in I}$.

Let I be an index set with $m + 1$ elements (that is $|I| = m + 1$). The points $\{x^i\}_{i \in I} \subset \mathbb{R}^n$ are called *affinely independent* if given any set $\{\lambda_i\}_{i \in I} \subset \mathbb{R}$ the conditions

$$\sum_{i \in I} \lambda_i x^i = 0 \qquad \text{and} \qquad \sum_{i \in I} \lambda_i = 0$$

imply that $\lambda_i = 0$ for any $i \in I$. As a consequence, the maximum number of affinely independent points in $\mathbb{R}^n$ is $n + 1$.

The convex hull of a collection $\{x^i\}_{i \in I}$ of affinely independent points is called the m-*simplex* spanned by $\{x^i\}_{i \in I}$. The elements x^i are called *vertices* of S. If $J \subseteq I$ with $|J| = k + 1$, the set $\mathrm{conv}\left(\{x^i\}_{i \in J}\right)$ is a k-*dimensional face* of S.

Notice that the vertices are 0-dimensional faces and S itself is also considered as a face of S. Moreover, any point $x \in S$ can be represented as a convex combination $x = \sum_{i \in I} \lambda_i x^i$ in a unique way. The coefficients $\{\lambda_i\}_{i \in I}$ are called the *barycentric coordinates* of x and it is possible to show that each barycentric coordinate λ_i depends continuously on x.

Let S be a m-simplex. A *simplicial subdivision* of S is a finite collection $\mathscr{P} = \{S_i\}_{i \in P}$ of m-simplexes (usually called *cells*) satisfying the following conditions:

- $\bigcup_{i \in P} S_i = S$,
- given any $i, j \in P$ the intersection $S_i \cap S_j$ either is empty or is a face of both S_i and S_j.

Clearly, given any $r > 0$ it is always possible to build a simplicial subdivision $\mathscr{P}$ of a m-simplex S such that

$$\mathrm{diam}(\mathscr{P}) = \max\{\mathrm{diam}(S') : S' \in \mathscr{P}\} < r.$$

Let $S = \mathrm{conv}\left(\{x^i\}_{i \in I}\right)$ be an m-simplex and $\mathscr{P}$ a simplicial subdivision. Denote by $V(\mathscr{P})$ the collection of all the vertices of all the cells in $\mathscr{P}$ (in particular each $x^i \in V(\mathscr{P})$). A *Sperner proper labeling* for the pair S and $\mathscr{P}$ is a function $v : V(\mathscr{P}) \to I$ such that $v(x) \in J$ for any $J \subseteq I$ with $x \in \mathrm{conv}\left(\{x^i\}_{i \in J}\right)$. A cell $S' \in \mathscr{P}$ is said to be *completely labeled* if $v(x) \neq v(y)$ holds for any pair of different vertices of S'.

Theorem A.1.1 (Sperner) *Let S be a simplex, $\mathscr{P}$ a simplicial subdivision, v a Sperner proper labeling associated with S and $\mathscr{P}$. Then, there exists an odd number of completely labeled cells.*

Fig. A.1 provides an example for the above result.

A.2 Functions

A function $g : \mathbb{R}^n \to \mathbb{R}$ is *lower semicontinuous* if one of the following equivalent conditions is satisfied:

- any $x \in \mathbb{R}^n$ satisfies

$$\liminf_{y \to x} g(y) \geq g(x);$$

- given any $x \in \mathbb{R}^n$ and $\epsilon > 0$ there exists a neighborhood U of x such that $g(y) \geq g(x) - \epsilon$ for all $y \in U$ with $y \neq x$;
- given any $a \in \mathbb{R}$ the sublevel set $\{x \in \mathbb{R}^n : g(x) \leq a\}$ is closed;
- the epigraph epi $g = \{(x, y) \in \mathbb{R}^n \times \mathbb{R} : y \geq g(x)\}$ is closed.

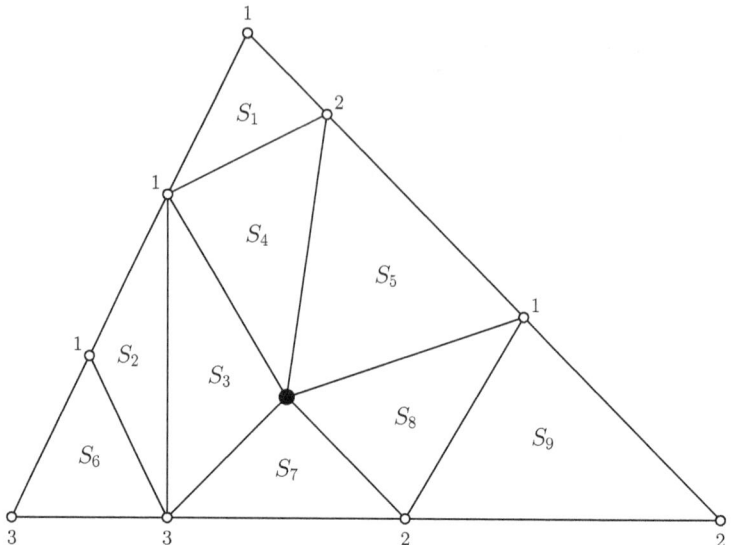

Fig. A.1 Whatever label the black point is given, at least one completely labeled cell exists

The function g is *upper semicontinuous* if $-g$ is lower semicontinuous and it is *continuous* if it is both lower and upper semicontinuous, i.e., given any $x \in \mathbb{R}^n$ and $\{x^k\}$ converging to x, then $g(x^k) \to g(x)$ holds. This fact implies that the graph of a continuous function is a closed set in $\mathbb{R}^n \times \mathbb{R}$.

A few of lower semicontinuity-preserving rules are listed below:

- if g is lower semicontinuous and $\alpha \geq 0$, then αg is lower semicontinuous,
- if g_1, g_2 are lower semicontinuous, then $g_1 + g_2$ is lower semicontinuous,
- if g_i are lower semicontinuous for any $i \in I$, where I is an arbitrary set of indices, and $g(x) = \sup_{i \in I} g_i(x)$ is finite for any $x \in \mathbb{R}^n$, then g is lower semicontinuous.

Several important results may be established under lower semicontinuity: one of the most important is the Weierstrass Theorem.

Theorem A.2.1 (Weierstrass) *Suppose $X \subseteq \mathbb{R}^n$ is compact and g is a lower semicontinuous function. Then, the set*

$$\arg\min\{g(z) : z \in X\} = \{x \in X \ : \ g(x) \leq g(z), \ \forall z \in X\}$$

is nonempty and closed.

A Weierstrass type result is possible even if the feasible region is not bounded but only closed under an additional property on g. The function g is called *coercive* if

$$\lim_{\|x\| \to +\infty} g(x) = +\infty.$$

The coercivity of g means that, given any constant M, there exists $r > 0$ such that $g(x) > M$ whenever $\|x\| > r$, or equivalently that the sublevel set $\{x \in \mathbb{R}^n : g(x) \leq a\}$ is bounded for any $a \in \mathbb{R}$.

Theorem A.2.2 *Suppose $X \subseteq \mathbb{R}^n$ is closed and g is a lower semicontinuous and coercive function. Then, $\arg\min\{g(z) : z \in X\}$ is a nonempty and closed set.*

The function g is said to be *Lipschitz continuous* if there is $L \geq 0$ such that

$$|g(x) - g(y)| \leq L\|x - y\|$$

holds for any $x, y \in \mathbb{R}^n$. If $L \leq 1$, the function g is called *nonexpansive*, while it is refereed to as a *contraction* if $L < 1$. Roughly speaking, the continuity of g means that small changes in x cause only small changes in $g(x)$, while its Lipschitz continuity allows checking and estimating these changes.

Now, the main concepts of differentiability are recalled. Given $x, d \in \mathbb{R}^n$, the function g is said to have *one-sided directional derivative* at x in the direction d if

$$g'(x; d) = \lim_{t \to 0^+} \frac{g(x + td) - g(x)}{t}$$

exists and it is finite. The function g is said to be *directionally differentiable* at x if the one-sided directional derivatives exist for every d and $g'(x; d) = -g'(x; -d)$. If g is directionally differentiable and e^i denotes the i-th vector of the canonical basis of $\mathbb{R}^n$, then $g'(x; e^i)$ is called *i-th partial derivative* of g at x and it is denoted by $\frac{\partial g}{\partial x_i}(x)$. The function g is said to be *differentiable* at x if there exists a vector $\nabla g(x)$, called *gradient*, such that

$$\lim_{\|d\| \to 0} \frac{|g(x + d) - g(x) - \langle \nabla g(x), d \rangle|}{\|d\|} = 0.$$

The following result provides an interesting relation between differentiability and directional differentiability.

Theorem A.2.3 *Let $x \in \mathbb{R}^n$ and $g : \mathbb{R}^n \to \mathbb{R}$ be given.*

a) If g is differentiable at x, then g is directionally differentiable and

$$g'(x; d) = \langle \nabla g(x), d \rangle$$

for any choice of $d \in \mathbb{R}^n$. Moreover, its gradient is given by

$$\nabla g(x) = \left(\frac{\partial g}{\partial x_1}(x), \ldots, \frac{\partial g}{\partial x_n}(x) \right).$$

b) If all the partial derivatives of g exist and are continuous at x, then g is differentiable at x.

The function g is said to be *continuously differentiable* when the partial derivatives of g exist and are continuous.

Now, consider the nonlinear minimization problem

$$\min\{f(x) : g_i(x) \leq 0,\ i \in I\} \qquad\qquad (P)$$

where $f : \mathbb{R}^n \to \mathbb{R}$ is the *objective function* and $g_i : \mathbb{R}^n \to \mathbb{R}$, with $i \in I = \{1, \ldots, m\}$, provide inequality constraints. A point $x \in \mathbb{R}^n$ is *feasible* if it satisfies the constraints, and the set of all feasible x is called the *feasible region*. The *index set of active constraints* at any feasible point x is denoted by

$$I(x) = \{i \in I : g_i(x) = 0\}.$$

If $i \notin I(x)$, then the i-th constraint is called inactive at x. A feasible point $\bar{x}$ is called a *local solution* of (P) if $f(x) \geq f(\bar{x})$ holds for any feasible x close to $\bar{x}$. The following necessary optimality conditions are called Karush–Kuhn–Tucker (KKT) conditions.

Theorem A.2.4 (KKT Conditions) *Let $\bar{x}$ be a local solution of (P) where f and all g_i are continuously differentiable. If the gradients $\nabla g_i(\bar{x})$, $i \in I(\bar{x})$, are linearly independent, then there exists a unique vector $\lambda = (\lambda_1, \ldots, \lambda_m) \in \mathbb{R}^m$ with $\lambda_i \geq 0$ for any $i \in I$ satisfying*

- *the Lagrange multipliers' rule*

$$\nabla f(\bar{x}) + \sum_{i \in I} \lambda_i \nabla g_i(\bar{x}) = 0$$

- *the complementary slackness conditions*

$$\lambda_i g_i(\bar{x}) = 0, \qquad \forall i \in I.$$

The necessary KKT conditions are also sufficient if (P) is a convex minimization problem, i.e., the objective function f and all the constraining functions g_i are convex functions.

Consider the parametric optimization problem

$$\min_{x}\{f(x, \varepsilon) : g_i(x, \varepsilon) \leq 0,\ i \in I\},$$

where $f : \mathbb{R}^n \times \mathbb{R}^p \to \mathbb{R}$ and $g_i : \mathbb{R}^n \times \mathbb{R}^p \to \mathbb{R}$, with $i \in I$, are continuously differentiable.

Theorem A.2.5 *Suppose that $f(\cdot, \varepsilon)$ is τ-convex with $\tau > 0$ and all $g_i(\cdot, \varepsilon)$ are convex for any $\varepsilon \in U$, where U is a neighborhood of a given $\bar{\varepsilon} \in \mathbb{R}^p$. Then, there exists a unique optimal solution $\bar{x}(\varepsilon)$ for any $\varepsilon \in U$. Moreover, suppose there exists*

$z(\varepsilon) \in \mathbb{R}^n$ *such that* $g_i(z(\varepsilon), \varepsilon) < 0$ *for any* i *and the gradients* $\nabla_x g_i(\bar{x}(\varepsilon), \varepsilon)$, $i \in I(\bar{x}(\varepsilon))$ *are linearly independent for any* $\varepsilon \in U$. *Then, there exists a unique Lagrange multipliers* $\bar{\lambda}(\varepsilon)$ *corresponding to* $\bar{x}(\varepsilon)$ *for any* $\varepsilon \in U$ *and the functions* $\bar{x}(\cdot)$ *and* $\bar{\lambda}(\cdot)$ *are continuous on* U.

Another important result is the so-called Mean Value Theorem.

Theorem A.2.6 *Let* $g : \mathbb{R}^n \to \mathbb{R}$ *be a differentiable function. Then, given any* $x, y \in \mathbb{R}^n$ *there exists* $t \in [0, 1]$ *such that*

$$g(y) - g(x) = \langle \nabla g(tx + (1-t)y), y - x \rangle$$

A *map* (or vector-valued function) $G : \mathbb{R}^n \to \mathbb{R}^m$ is a function whose values $G(x) = (g_1(x), \ldots, g_m(x))$ are vectors in $\mathbb{R}^m$ with components $g_i : \mathbb{R}^n \to \mathbb{R}$. It is called continuous/Lipschitz/differentiable if each component function g_i is continuous/Lipschitz/differentiable. In particular, the differentiability of G can be equivalently defined as

$$\lim_{\|d\| \to 0} \frac{\|G(x+d) - G(x) - \nabla G(x)d\|}{\|d\|} = 0.$$

where

$$\nabla G(x) = \begin{pmatrix} \frac{\partial g_1}{\partial x_1}(x) & \frac{\partial g_1}{\partial x_2}(x) & \ldots & \frac{\partial g_1}{\partial x_n}(x) \\ \frac{\partial g_2}{\partial x_1}(x) & \frac{\partial g_2}{\partial x_2}(x) & \ldots & \frac{\partial g_2}{\partial x_n}(x) \\ \vdots & \vdots & \ddots & \vdots \\ \frac{\partial g_m}{\partial x_1}(x) & \frac{\partial g_m}{\partial x_2}(x) & \ldots & \frac{\partial g_m}{\partial x_n}(x) \end{pmatrix}$$

is a $m \times n$ matrix, called *Jacobian matrix*, such that its i-th row is the gradient of the i-th component g_i. Let $g : \mathbb{R}^n \to \mathbb{R}$ be a differentiable function. If the gradient ∇g is differentiable, that is all of its components are differentiable, the Jacobian matrix of ∇g at a point x is called *Hessian matrix* and it is denoted by $\nabla^2 g(x)$. The components of $\nabla^2 g(x)$ are the second partial derivatives of g and the Hessian matrix reads

$$\nabla^2 g(x) = \begin{pmatrix} \frac{\partial^2 g}{\partial x_1^2}(x) & \frac{\partial^2 g}{\partial x_1 \partial x_2}(x) & \ldots & \frac{\partial^2 g}{\partial x_1 \partial x_n}(x) \\ \frac{\partial^2 g}{\partial x_2 \partial x_1}(x) & \frac{\partial^2 g}{\partial x_2^2}(x) & \ldots & \frac{\partial^2 g}{\partial x_2 \partial x_n}(x) \\ \vdots & \vdots & \ddots & \vdots \\ \frac{\partial^2 g}{\partial x_n \partial x_1}(x) & \frac{\partial^2 g}{\partial x_n \partial x_2}(x) & \ldots & \frac{\partial^2 g}{\partial x_n^2}(x) \end{pmatrix}$$

The mixed derivatives of g are the entries off the main diagonal in the Hessian matrix. Supposing that they are continuous, i.e., g is twice continuously differentiable, the order of differentiation does not matter, that is

$$\frac{\partial^2 g}{\partial x_j \partial x_i} = \frac{\partial}{\partial x_i}\left(\frac{\partial g}{\partial x_j}\right) = \frac{\partial}{\partial x_j}\left(\frac{\partial g}{\partial x_i}\right) = \frac{\partial^2 g}{\partial x_i \partial x_j},$$

and the Hessian matrix is symmetric.

A point $x \in \mathbb{R}^n$ is a *fixed point* of the map $G : \mathbb{R}^n \to \mathbb{R}^n$ if $G(x) = x$. The set of the fixed points of G is denoted by fix G. One of the most famous fixed point theorems for continuous functions was proved by Brouwer and it has been used across numerous fields of mathematics.

Theorem A.2.7 (Brouwer) *Suppose $G : \mathbb{R}^n \to \mathbb{R}^n$ is continuous and $C \subseteq \mathbb{R}^n$ is a nonempty convex compact set such that $G(x) \in C$ for any $x \in C$. Then, there exists at least one $x \in C$ such that $x = G(x)$.*

A.3 Multivalued Maps

A *multivalued map* $\Phi : \mathbb{R}^n \rightrightarrows \mathbb{R}^m$ assigns to each $x \in \mathbb{R}^n$ a possibly empty subset $\Phi(x)$ of $\mathbb{R}^m$. The *domain* of Φ is

$$\text{dom}\,\Phi = \{x \in \mathbb{R}^n : \Phi(x) \neq \emptyset\},$$

the *image* of a set $X \subseteq \mathbb{R}^n$ under Φ is the set

$$\Phi(X) = \bigcup_{x \in X} \Phi(x)$$

and the *range* of Φ is the image of $\mathbb{R}^n$. Moreover, the multivalued map Φ can be identified with its *graph*, which is the subset of $\mathbb{R}^n \times \mathbb{R}^m$ given by

$$\text{gph}\,\Phi = \{(x, y) \in \mathbb{R}^n \times \mathbb{R}^m : y \in \Phi(x)\}.$$

A multivalued map $\Phi : \mathbb{R}^n \rightrightarrows \mathbb{R}^m$ has a natural inverse $\Phi^{-1} : \mathbb{R}^m \rightrightarrows \mathbb{R}^n$ defined by

$$\Phi^{-1}(y) = \{x \in \mathbb{R}^n : y \in \Phi(x)\}.$$

The set $\Phi^{-1}(y)$ is called the *lower section* of Φ at y. The multivalued map $\Phi : \mathbb{R}^n \rightrightarrows \mathbb{R}^m$ is said to be closed-valued if $\Phi(x)$ is a closed set for any $x \in \mathbb{R}^n$. The terms open-valued, compact-valued, convex-valued and so on are similarly defined.

Recall that a map $G : \mathbb{R}^n \to \mathbb{R}^m$ is continuous at $x \in \mathbb{R}^n$ if and only if $G(x^k) \to G(x)$ whenever $\{x^k\} \to x$. This fact is equivalent to state that given any open set Ω such that $G(x) \in \Omega$ there exists a neighborhood U of x such that $G(z) \in \Omega$ for any $z \in U$. Note that the inclusion $G(x) \in \Omega$ can be generalized in two different ways when $G = \Phi$ is a multivalued map: either $\Phi(x) \subseteq \Omega$ or $\Phi(x) \cap \Omega \neq \emptyset$. This leads to two different notions of continuity for multivalued maps.

A multivalued map $\Phi : \mathbb{R}^n \rightrightarrows \mathbb{R}^m$ is said to be

- *upper semicontinuous* at x if given any open set Ω for which $\Phi(x) \subseteq \Omega$ there exists a neighborhood U of x such that $\Phi(z) \subseteq \Omega$ holds for any $z \in U$;
- *lower semicontinuous* at x if for given any open set Ω for which $\Phi(x) \cap \Omega \neq \emptyset$ there exists a neighborhood U of x such that $\Phi(z) \cap \Omega \neq \emptyset$ holds for any $z \in U$.

The map Φ is called upper (respectively lower) semicontinuous if it is upper (lower) semicontinuous at every point of dom Φ.

If Φ is single-valued (that is, Φ is a map with dom $\Phi = \mathbb{R}^n$), both definitions coincide with the definition of continuity of a map. However, the following examples show that these two definitions do not coincide when Φ is multivalued.

Example A.3.1 The multivalued map $\Phi : \mathbb{R} \rightrightarrows \mathbb{R}$ defined by

$$\Phi(x) = \begin{cases} \{0\} & \text{if } x \neq 0, \\ (-1, 1) & \text{if } x = 0, \end{cases}$$

is upper semicontinuous everywhere but it is not lower semicontinuous at $x = 0$. Indeed, the open set $\Omega = (0, 1)$ satisfies $\Phi(0) \cap \Omega \neq \emptyset$ and $\Phi(z) \cap \Omega = \emptyset$ for any $z \neq 0$ (see Fig. A.2).

Example A.3.2 The multivalued map $\Phi : \mathbb{R} \rightrightarrows \mathbb{R}$ defined by

$$\Phi(x) = \begin{cases} [-1, 1] & \text{if } x \neq 0, \\ \{0\} & \text{if } x = 0, \end{cases}$$

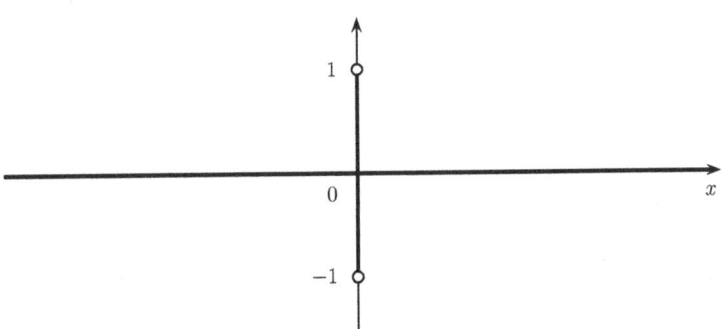

Fig. A.2 Graph of the multivalued map of Example A.3.1

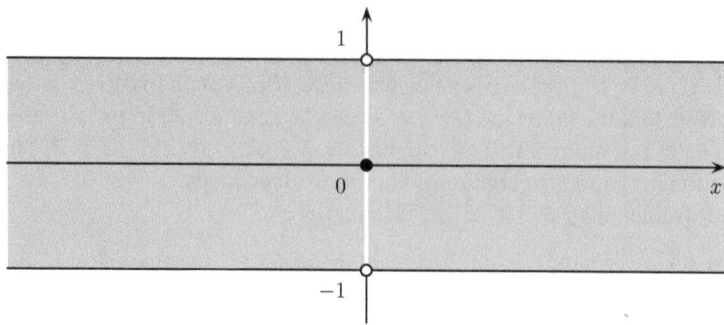

Fig. A.3 Graph of the multivalued map of Example A.3.2

is lower semicontinuous everywhere but it is not upper semicontinuous at $x = 0$. Indeed, the open set $\Omega = (-1/2, 1/2)$ satisfies $\Phi(0) \subset \Omega$ and $\Phi(z) \not\subseteq \Omega$ for any $z \neq 0$ (see Fig. A.3).

Although both concepts coincide with the notion of continuity in the single-valued case, the previous examples shows that neither implies the closedness of gph Φ.

A multivalued map $\Phi : \mathbb{R}^n \rightrightarrows \mathbb{R}^m$ is said to be *closed* if $y \in \Phi(x)$ for any sequence $\{(x^k, y^k)\} \subseteq$ gph Φ converging to any (x, y), i.e., gph Φ is a closed set of $\mathbb{R}^n \times \mathbb{R}^m$.

The following example shows that the closedness is not related to the previous concepts of semicontinuity.

Example A.3.3 The multivalued map $\Phi : \mathbb{R} \rightrightarrows \mathbb{R}$ defined by

$$\Phi(x) = \begin{cases} \{0\} & \text{if } x = 0, \\ [1/|x|, +\infty) & \text{if } x \neq 0, \end{cases}$$

is closed, but it is not both upper and lower semicontinuous at $x = 0$. Indeed, the open set $\Omega = (-1, 1)$ satisfies $\Phi(0) \subset \Omega$ and $\Phi(z) \cap \Omega = \emptyset$ for any $z \in (-1, 0) \cup (0, 1)$ (see Fig. A.4).

Closed multivalued map are always closed-valued. The converse is false as Example A.3.2 shows. However, the Closed Graph Theorem highlights the strict relation between upper semicontinuity and closedness of a multivalued map.

Theorem A.3.4 *An upper semicontinuous multivalued map Φ with closed values is closed. The converse holds if range of Φ is bounded.*

The Closed Graph Theorem can be view as a characterization of the upper semicontinuity in terms of sequences under suitable assumptions. Another useful characterization of the upper semicontinuity in terms of sequences is the following.

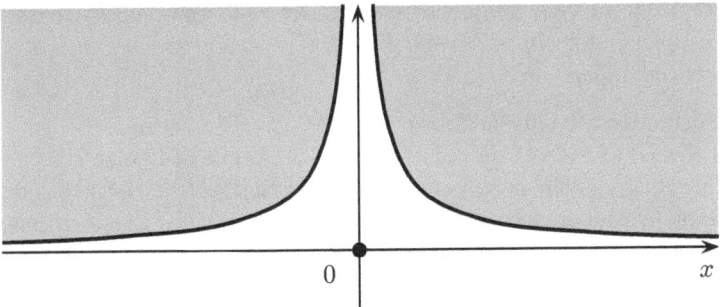

Fig. A.4 Graph of the multivalued map of Example A.3.3

Theorem A.3.5 *A compact-valued multivalued map Φ is upper semicontinuous if and only if given any sequence $\{x^k\}$ converging to x and any $\{y^k\}$ with $y^k \in \Phi(x^k)$ there exists a subsequence $\{y^{k_\ell}\}$ converging to some $y \in \Phi(x)$.*

A characterization of lower semicontinuity for a multivalued map holds without any assumption of compactness.

Theorem A.3.6 *Given a multivalued map Φ, the following statements are equivalent:*

a) Φ is lower semicontinuous;
b) given any sequence $\{x^k\}$ converging to x and any $y \in \Phi(x)$ there exists a subsequence $\{x^{k_\ell}\}$ of $\{x^k\}$ and a sequence $\{y^{k_\ell}\}$ converging to y with $y^{k_\ell} \in \Phi(x^{k_\ell})$ for all ℓ.

If dom $\Phi = \mathbb{R}^n$, *then a) and b) are equivalent to the following statement:*

c) given any sequence $\{x^k\}$ converging to x and any $y \in \Phi(x)$ there exists a sequence $\{y^k\}$ converging to y with $y^k \in \Phi(x^k)$ for all k.

A relation between lower semicontinuity and openness of the graph can be stated in a manner similar to the Closed Graph Theorem A.3.4.

Theorem A.3.7 *Suppose the multivalued map Φ is open and convex-valued, then* gph Φ *is open in* dom $\Phi \times \mathbb{R}^n$ *if and only if Φ is lower semicontinuous.*

Finally, two interesting results provide sufficient conditions for the lower semicontinuity of a multivalued map.

Theorem A.3.8 *If the multivalued map Φ has open lower sections, then it is lower semicontinuous.*

Theorem A.3.9 *Let Φ_1, Φ_2 be two multivalued maps with* dom $\Phi_1 = $ dom Φ_2. *If* gph Φ_1 *is open in* dom $\Phi_1 \times \mathbb{R}^m$ *and Φ_2 is lower semicontinuous, then $\Phi_1 \cap \Phi_2$ is lower semicontinuous.*

A selection of a multivalued map $\Phi : \mathbb{R}^n \rightrightarrows \mathbb{R}^m$ is a map $G : \mathbb{R}^n \rightarrow \mathbb{R}^m$ that satisfies $G(x) \in \Phi(x)$ for each $x \in \mathbb{R}^n$. The Axiom of Choice guarantees that multivalued maps with nonempty values always admit selections, but they may have no additional useful properties. The following result is one of a series of interesting results on the existence of continuous selections given by Michael.

Theorem A.3.10 (Michael)   *Assume that the multivalued map Φ is lower semicontinuous with* dom $\Phi \neq \emptyset$ *and convex values, then it admits a continuous selection.*

A point $x \in \mathbb{R}^n$ is a fixed point of the multivalued map $\Phi : \mathbb{R}^n \rightrightarrows \mathbb{R}^n$ if $x \in \Phi(x)$. The set of the fixed points of Φ is denoted by fix Φ. The Brouwer fixed point Theorem A.2.7 and the Michael selection Theorem A.3.10 lead to the following fixed point result for lower semicontinuous multivalued maps.

Theorem A.3.11 *Suppose $\Phi : \mathbb{R}^n \rightrightarrows \mathbb{R}^n$ is lower semicontinuous with nonempty convex values and $C \subseteq \mathbb{R}^n$ is a nonempty convex compact set such that $\Phi(x) \subseteq C$ for each $x \in C$. Then there exists at least one $x \in C$ such that $x \in \Phi(x)$.*

References

1. M. Ait Mansour, H. Riahi, Sensitivity analysis for abstract equilibrium problems. J. Math. Anal. Appl. **306**, 684–691 (2005)
2. B. Alleche, V.D. Rădulescu, M. Sebaoui, The Tikhonov regularization for equilibrium problems and applications to quasi-hemivariational inequalities. Optim. Lett. **9**, 483–503 (2015)
3. L.Q. Anh, P.Q. Khanh, Hölder continuity of the unique solution to quasiequilibrium problems in metric spaces. J. Optim. Theory Appl. **141**, 37–54 (2009)
4. L.Q. Anh, P.Q. Khanh, T.N. Tam, On Hölder continuity of approximate solutions to parametric equilibrium problems. Nonlinear Anal. **75**, 2293–2303 (2009)
5. L.Q. Anh, A.Y. Kruger, N.H. Thao, On Hölder calmness of solution mappings in parametric equilibrium problems. Top **22**, 331–342 (2014)
6. P.N. Anh, An LQ regularization method for pseudomonotone equilibrium problems on polyhedra. Vietnam J. Math. **36**, 209–228 (2008)
7. P.N. Anh, A logarithmic quadratic regularization method for solving pseudomonotone equilibrium problems. Acta Math. Vietnam. **34**, 183–200 (2009)
8. P.N. Anh, H.A. Le Thi, An Armijo-type method for pseudomonotone equilibrium problems and its applications. J. Glob. Optim. **57** 803–820 (2013)
9. P.N. Anh, H.A. Le Thi, The subgradient extragradient method extended to equilibrium problems. Optimization **64**, 225–248 (2015)
10. P.N. Anh, T.N. Hai, P.M. Tuan, On ergodic algorithms for equilibrium problems. J. Glob. Optim. **64**, 179–195 (2016)
11. P.N. Anh, T.T.H. Anh, N.D. Hien, Modified basic projection methods for a class of equilibrium problems. Numer. Algorithms **79**, 139–152 (2018)
12. A.S. Antipin, Inverse problems of nonlinear programming. Comput. Math. Model. **7**, 263–287 (1996)
13. K.J. Arrow, G. Debreu, Existence of an equilibrium for a competitive economy. Econometrica **22**, 265–290 (1954)
14. A. Auslender, *Optimisation: Méthodes Numériques* (Masson, Paris, 1976)
15. D. Aussel, N. Hadjisavvas, On quasimonotone variational inequalities. J. Optim. Theory Appl. **121**, 445–450 (2004)
16. D. Aussel, J. Cotrina, A.N. Iusem, An existence result for quasi-equilibrium problems. J. Convex Anal. **24**, 55–66 (2017)
17. C. Baiocchi, A. Capelo, *Variational and Quasivariational Inequalities: Applications to Free Boundary Problems* (Wiley, New York, 1984)

© Springer Nature Switzerland AG 2019
G. Bigi et al., *Nonlinear Programming Techniques for Equilibria*, EURO Advanced Tutorials on Operational Research, https://doi.org/10.1007/978-3-030-00205-3

18. M.L. Balinski, W.J. Baumol, The dual in nonlinear programming and its economic interpretation. Rev. Econ. Stud. **35**, 237–256 (1968)
19. H.H. Bauschke, J.M. Borwein, On projection algorithms for solving convex feasibility problems. SIAM Rev. **38**, 367–426 (1996)
20. J.Y. Bello Cruz, P.S.M. Santos, S. Scheimberg, A two-phase algorithm for a variational inequality formulation of equilibrium problems. J. Optim. Theory Appl. **159**, 562–575 (2013)
21. A. Bensoussan, J.-L. Lions, Nouvelle formulation de problèmes de contrôle impulsionnel et applications. C. R. Hebd. Séances Acad. Sci. Sér. A **276**, 1189–1192 (1973)
22. A. Bensoussan, M. Goursat, J.-L. Lions, Contrôle impulsionnel et inéquations quasi-variationnelles stationnaires. C. R. Hebd. Séances Acad. Sci. Sér. A **276**, 1279–1284 (1973)
23. C. Berge, *Espaces topologiques et fonctions multivoques* (Dunod, Paris, 1959)
24. M. Bianchi, R. Pini, Coercivity conditions for equilibrium problems. J. Optim. Theory Appl. **124**, 79–92 (2005)
25. M. Bianchi, S. Schaible, Generalized monotone bifunctions and equilibrium problems. J. Optim. Theory Appl. **90**, 31–43 (1996)
26. M. Bianchi, G. Kassay, R. Pini, Existence of equilibria via Ekeland's principle. J. Math. Anal. Appl. **305**, 502–512 (2005)
27. M. Bianchi, G. Kassay, R. Pini, Well-posed equilibrium problems. Nonlinear Anal. **72**, 460–468 (2010)
28. G. Bigi, M. Passacantando, Gap functions and penalization for solving equilibrium problems with nonlinear constraints. Comput. Optim. Appl. **53**, 323–346 (2012)
29. G. Bigi, M. Passacantando, D-gap functions and descent techniques for solving equilibrium problems. J. Glob. Optim. **62**, 183–203 (2015)
30. G. Bigi, M. Passacantando, Descent and penalization techniques for equilibrium problems with nonlinear constraints. J. Optim. Theory Appl. **164**, 804–818 (2015)
31. G. Bigi, M. Passacantando, Twelve monotonicity conditions arising from algorithms for equilibrium problems. Optim. Methods Softw. **30**, 323–337 (2015)
32. G. Bigi, M. Passacantando, Gap functions for quasi-equilibria. J. Glob. Optim. **66**, 791–810 (2016)
33. G. Bigi, M. Passacantando, Auxiliary problem principles for equilibria. Optimization **66**, 1955–1972 (2017)
34. G. Bigi, M. Passacantando, Differentiated oligopolistic markets with concave cost functions via Ky Fan inequalities. Decis. Econ. Finance **40**, 63–79 (2017)
35. G. Bigi, M. Passacantando, Convergence of the fixed point algorithm for quasi-equilibria. Technical Report Dipartimento di Informatica, Universitáá di Pisa, http://eprints.adm.unipi.it/2389/ (2018)
36. G. Bigi, M. Castellani, M. Pappalardo, A new solution method for equilibrium problems. Optim. Methods Softw. **24**, 895–911 (2009)
37. S.C. Billups, K.G. Murty, Complementarity problems. J. Comput. Appl. Math. **124**, 303–318 (2000)
38. E. Blum, W. Oettli, From optimization and variational inequalities to equilibrium problems. Math. Student **63**, 123–145 (1994)
39. M. Bogdan, J. Kolumban, Some regularities for parametric equilibrium problems. J. Glob. Optim. **44**, 481–492 (2009)
40. J.F. Bonnans, A. Shapiro, *Perturbation Analysis of Optimization Problems* (Springer, New York, 2000)
41. K.C. Border, *Fixed Point Theorems with Applications to Economics and Game Theory* (Cambridge University Press, Cambridge, 1985)
42. R. Burachik, G. Kassay, On a generalized proximal point method for solving equilibrium problems in Banach spaces. Nonlinear Anal. **75**, 6456–6464 (2012)
43. H. Brézis, L. Nirenberg, G. Stampacchia, A remark on Ky Fan's minimax principle. Boll. Unione Mat. Ital. **6**, 293–300 (1972)
44. M. Castellani, M. Giuli, On equivalent equilibrium problems. J. Optim. Theory Appl. **147**, 157–168 (2010)

45. M. Castellani, M. Giuli, Refinements of existence results for relaxed quasimonotone equilibrium problems. J. Glob. Optim. **57**, 1213–1227 (2103)
46. M. Castellani, M. Giuli, Ekeland's principle for cyclically antimonotone equilibrium problems. Nonlinear Anal. Real World Appl. **32**, 213–228 (2016)
47. M. Castellani, M. Pappalardo, Characterizations of ρ-convex functions, in *Generalized Convexity, Generalized Monotonicity: Recent Results*. Nonconvex Optimization and Its Applications, vol. 27 (Kluwer Academic, Dordrecht, 1998), pp. 219–233
48. M. Castellani, M. Pappalardo, M. Passacantando, Existence results for nonconvex equilibrium problems. Optim. Methods Softw. **25**, 49–58 (2010)
49. M. Castellani, M. Giuli, M. Pappalardo, A Ky Fan minimax inequality for quasiequilibria on finite-dimensional spaces. J. Optim. Theory Appl. **179**, 53–64 (2018)
50. O. Chadli, I.V. Konnov, J.-C. Yao, Descent methods for equilibrium problems in a Banach space. Comput. Math. Appl. **48**, 609–616 (2004)
51. D. Chan, J.-S. Pang, The generalized quasi-variational inequality problem. Math. Oper. Res. **7**, 211–222 (1982)
52. Ch. Charita, A note on D-gap functions for equilibrium problems. Optimization **62**, 211–226 (2013)
53. P. Cubiotti, Existence of solutions for lower semicontinuous quasi-equilibrium problems. Comput. Math. Appl. **30**, 11–22 (1995)
54. S.C. Dafermos, Traffic equilibrium and variational inequalities. Transp. Sci. **14**, 42–54 (1980)
55. J.M. Danskin, *The Theory of Max-Min and Its Applications to Weapons Allocation Problems* (Springer, New York, 1967)
56. G. Debreu, A social equilibrium existence theorem. Proc. Natl. Acad. Sci. U S A **38**, 886–893 (1952)
57. D. Di Lorenzo, M. Passacantando, M. Sciandrone, A convergent inexact solution method for equilibrium problems. Optim. Methods Softw. **29**, 979–991 (2014)
58. B.V. Dinh, D.S. Kim, Projection algorithms for solving nonmonotone equilibrium problems in Hilbert space. J. Comput. Appl. Math. **302**, 106–117 (2016)
59. B.V. Dinh, L.D. Muu, A projection algorithm for solving pseudomonotone equilibrium problems and its application to a class of bilevel equilibria. Optimization **64**, 559–575 (2015)
60. B.V. Dinh, P.G. Hung, L.D. Muu, Bilevel optimization as a regularization approach to pseudomonotone equilibrium problems. Numer. Funct. Anal. Optim. **35**, 539–563 (2014)
61. A. Dreves, F. Facchinei, C. Kanzow, S. Sagratella, On the solution of the KKT conditions of generalized Nash equilibrium problems. SIAM J. Optim. **21**, 1082–1108 (2011)
62. F. Facchinei, C. Kanzow, Generalized Nash equilibrium problems. Ann. Oper. Res. **175**, 177–211 (2010)
63. F. Facchinei, A. Fischer, V. Piccialli, Generalized Nash equilibrium problems and Newton methods. Math. Program. **117**, 163–194 (2009)
64. F. Facchinei, C. Kanzow, S. Sagratella, Solving quasi-variational inequalities via their KKT conditions. Math. Program. **144**, 369–412 (2014)
65. K. Fan, A generalization of Tychonoff's fixed point theorem. Math. Ann. **142**, 305–310 (1961)
66. K. Fan, A minimax inequality and applications, in *Inequalities III*, ed. by O. Shisha (Academic, New York, 1972), pp. 103–113
67. S.D. Flåm, A.S. Antipin, Equilibrium programming using proximal-like algorithms. Math. Program. **78**, 29–41 (1997)
68. M. Fukushima, Equivalent differentiable optimization problems and descent methods for asymmetric variational inequality problems. Math. Program. **53**, 99–110 (1992)
69. Z. Han, D. Niyato, W. Saad, T. Basar, A. Hjorungnes, *Game Theory in Wireless and Communication Networks* (Cambridge University Press, New York, 2012)
70. D.V. Hieu, A new shrinking gradient-like projection method for equilibrium problems. Optimization **66**, 2291–2307 (2017)
71. W.W. Hogan, Point-to-set maps in mathematical programming. SIAM Rev. **15**, 591–603 (1973)

72. P.G. Hung, L.D. Muu, The Tikhonov regularization extended to equilibrium problems involving pseudomonotone bifunctions. Nonlinear Anal. **74**, 6121–6129 (2011)
73. A.N. Iusem, W. Sosa, Iterative algorithms for equilibrium problems. Optimization **52**, 301–316 (2003)
74. A.F. Izmailov, M.V. Solodov, On error bounds and Newton-type methods for generalized Nash equilibrium problems. Comput. Optim. Appl. **59**, 201–218 (2014)
75. B. Knaster, C. Kuratowski, S. Mazurkiewicz, Ein Beweies des Fixpunktsatzes für N Dimensionale Simplexe. Fundam. Math. **14**, 132–137 (1929)
76. S. Komlósi, Generalized monotonicity and generalized convexity. J. Optim. Theory Appl. **84**, 361–376 (1995)
77. I.V. Konnov, Combined relaxation method for monotone equilibrium problems. J. Optim. Theory Appl. **111**, 327–340 (2001)
78. I.V. Konnov, Application of the proximal point method to nonmonotone equilibrium problems. J. Optim. Theory Appl. **119**, 317–333 (2003)
79. I.V. Konnov, Regularization method for nonmonotone equilibrium problems. J. Nonlinear Convex Anal. **10**, 93–101 (2009)
80. I.V. Konnov, M.S.S. Ali, Descent methods for monotone equilibrium problems in Banach spaces. J. Comput. Appl. Math. **188**, 165–179 (2006)
81. I.V. Konnov, D.A. Dyabilkin, Nonmonotone equilibrium problems: coercivity conditions and weak regularization. J. Glob. Optim. **49**, 575–587 (2011)
82. I.V. Konnov, O.V. Pinyagina, Descent method with respect to the gap function for nonsmooth equilibrium problems. Russ. Math. **47**, 67–73 (2003)
83. I.V. Konnov, O.V. Pinyagina, D-gap functions for a class of equilibrium problems in Banach spaces. Comput. Methods Appl. Math. **3**, 274–286 (2003)
84. I.V. Konnov, O.V. Pinyagina, D-gap functions and descent methods for a class of monotone equilibrium problems. Lobachevskii J. Math. **13**, 57–65 (2003)
85. I.V. Konnov, S. Schaible, J.C. Yao, Combined relaxation method for mixed equilibrium problems. J. Optim. Theory Appl. **126**, 309–322 (2005)
86. K. Kubota, M. Fukushima, Gap function approach to the generalized Nash equilibrium problem. J. Optim. Theory Appl. **144**, 511–531 (2010)
87. N. Langenberg, Interior proximal methods for equilibrium programming: part I. Optimization **62**, 1247–1266 (2013)
88. N. Langenberg, Interior proximal methods for equilibrium programming: part II. Optimization **62**, 1603–1625 (2013)
89. W. Leontief, *The Structure of American Economy 1919–1929* (Oxford University Press, London, 1949)
90. H.M. Markowitz, Portfolio selection. J. Finance **7**, 77–91 (1952)
91. G. Mastroeni, On auxiliary principle for equilibrium problems, in *Equilibrium Problems and Variational Models*, ed. by P. Daniele, F. Giannessi, A. Maugeri (Kluwer Academic, Norwell, 2003), pp. 289–298
92. G. Mastroeni, Gap functions for equilibrium problems. J. Glob. Optim. **27**, 411–426 (2003)
93. G. Mastroeni, M. Pappalardo, M. Passacantando, Merit functions: a bridge between optimization and equilibria. 4 OR **12**, 1–33 (2014)
94. E. Michael, Continuous selections I. Ann. Math. **63**, 361–382 (1956)
95. G.J. Minty, On the generalization of a direct method of the calculus of variations. Bull. Am. Math. Soc. **73**, 315–321 (1967)
96. U. Mosco, Implicit variational problems and quasi variational inequalities, in *Nonlinear Operators and the Calculus of Variations*, ed. by J.P. Gossez, E.J. Lami Dozo, J. Mawhin, L. Waelbroeck (Springer, Berlin, 1976), pp. 83–156
97. A. Moudafi, Proximal point methods extended to equilibrium problems. J. Nat. Geom. **15**, 91–100 (1999)
98. A. Moudafi, M. Théra, Proximal and dynamical approaches to equilibrium problems, in *Ill-Posed Variational Problems and Regularization Techniques*, ed. by M. Théra, R. Tichatschke (Springer, Berlin, 1999), pp. 187–201

99. L.D. Muu, W. Oettli, Convergence of an adaptive penalty scheme for finding constrained equilibria. Nonlinear Anal. **18**, 1159–1166 (1992)

100. L.D. Muu, T.D. Quoc, Regularization algorithms for solving monotone Ky Fan inequalities with application to a Nash-Cournot equilibrium model. J. Optim. Theory Appl. **142**, 185–204 (2009)

101. Y. Nesterov, L. Scrimali, Solving strongly monotone variational and quasi-variational inequalities. Discrete Contin. Dyn. Syst. **31**, 1383–1396 (2011)

102. T.P.D. Nguyen, J.J. Strodiot, T.T.V. Nguyen, V.H. Nguyen, A family of extragradient methods for solving equilibrium problems. J. Ind. Manag. Optim. **11**, 619–630 (2015)

103. T.T.V. Nguyen, J.J. Strodiot, V.H. Nguyen, A bundle method for solving equilibrium problems. Math. Program. **116**, 529–552 (2009)

104. T.T.V. Nguyen, J.J. Strodiot, V.H. Nguyen, The interior proximal extragradient method for solving equilibrium problems. J. Glob. Optim. **44**, 175–192 (2009)

105. T.T.V. Nguyen, T.P.D. Nguyen, J.J. Strodiot, V.H. Nguyen, A class of hybrid methods for quasi-variational inequalities. Optim. Lett. **8**, 2211–2226 (2014)

106. H. Nikaido, K. Isoda, Note on noncooperative convex games. Pac. J. Math. **5**, 807–815 (1955)

107. J. Outrata, M. Kocvara, On a class of quasi-variational inequalities. Optim. Methods Softw. **5**, 275–295 (1995)

108. J. Outrata, J. Zowe, A Newton method for a class of quasi-variational inequalities. Comput. Optim. Appl. **4**, 5–21 (1995)

109. J.-S. Pang, M. Fukushima, Quasi-variational inequalities, generalized Nash equilibria, and multileader-follower games. Comput. Manag. Sci. **2**, 21–56 (2005)

110. M. Patriksson, *The Traffic Assignment Problem: Models and Methods* (VSP International Science, Utrecht, 1994)

111. J.-M. Peng, Equivalence of variational inequality problems to unconstrained minimization. Math. Program. **78**, 347–355 (1997)

112. T.D. Quoc, L.D Muu, Iterative methods for solving monotone equilibrium problems via dual gap functions. Comput. Optim. Appl. **51**, 709–728 (2012)

113. T.D. Quoc, M.L. Dung, V.H. Nguyen, Extragradient algorithms extended to equilibrium problems. Optimization **57**, 749–776 (2008)

114. T.D. Quoc, P.N. Anh, L.D. Muu, Dual extragradient algorithms extended to equilibrium problems. J. Glob. Optim. **52**, 139–159 (2012)

115. F.M.P. Raupp, W. Sosa, An analytic center cutting plane algorithm for finding equilibrium points. RAIRO Oper. Res. **40**, 37–52 (2006)

116. I.P. Ryazantseva, First-order methods for certain quasi-variational inequalities in Hilbert space. Comput. Math. Math. Phys. **47**, 183–190 (2007)

117. I.W. Sandberg, A nonlinear input-output model of a multisectored economy. Econometrica **41**, 1167–1182 (1973)

118. P.S.M. Santos, S. Scheimberg, An inexact subgradient algorithm for equilibrium problems. Comput. Appl. Math. **30**, 91–107 (2011)

119. P.S.M. Santos, S. Scheimberg, An outer approximation algorithm for equilibrium problems in Hilbert spaces. Optim. Methods Softw. **30**, 379–390 (2015)

120. S. Scheimberg, P.S.M. Santos, A relaxed projection method for finite-dimensional equilibrium problems. Optimization **60**, 1193–1208 (2011)

121. A.H. Siddiqi, Q.H. Ansari, Strongly nonlinear quasivariational inequalities. J. Math. Anal. Appl. **149**, 444–450 (1990)

122. M. Smith, Existence, uniqueness, and stability of traffic equilibria. Transp. Res. B **13**, 259–304 (1979)

123. E. Sperner, Neuer beweis für die invarianz der dimensionszahl und des gebietes. Abh. Math. Sem. Univ. Hamburg **6**, 265–272 (1928)

124. J.J. Strodiot, T.T.V. Nguyen, V.H. Nguyen, A new class of hybrid extragradient algorithms for solving quasi-equilibrium problems. J. Glob. Optim. **56**, 373–397 (2013)

125. J.J. Strodiot, P.T. Vuong, T.T.V. Nguyen, A class of shrinking projection extragradient methods for solving non-monotone equilibrium problems in Hilbert spaces. J. Glob. Optim. **64**, 159–178 (2016)

126. N.T.T. Van, J.J. Strodiot, V.H. Nguyen, P.T. Vuong, An extragradient-type method for solving nonmonotone quasi-equilibrium problems. Optimization **67**, 651–664 (2018)
127. J.P. Vial, Strong and weak convexity of sets and functions. Math. Oper. Res. **8**, 231–259 (1983)
128. J.G. Wardrop, Some theoretical aspects of road traffic research. Proc. Inst. Civ. Eng. **1**, 325–362 (1952)
129. N. Yamashita, K. Taji, M. Fukushima, Unconstrained optimization reformulations of variational inequality problems. J. Optim. Theory Appl. **92**, 439–456 (1997)
130. M. Yoseloff, Topologic proofs of some combinatorial theorems. J. Combin. Theory Ser. A **17**, 95–111 (1974)
131. J. Yu, H. Yang, C. Yu, Well posed Ky Fan's point, quasi-variational inequality and Nash equilibrium problems. Nonlinear Anal. **66**, 777–790 (2007)
132. L.C. Zeng, J.C. Yao, Modified combined relaxation method for general monotone equilibrium problems in Hilbert spaces. J. Optim. Theory Appl. **131**, 469–483 (2006)
133. J. Zhang, C. Xu, Inverse optimization for linearly constrained convex separable programming problems. Eur. J. Oper. Res. **200**, 671–679 (2010)
134. J. Zhang, B. Qu, N. Xiu, Some projection-like methods for the generalized Nash equilibria. Comput. Optim. Appl. **45**, 89–109 (2010)
135. L. Zhang, J.Y. Han, Unconstrained optimization reformulations of equilibrium problems. Acta Math. Sin. **25**, 343–354 (2009)
136. L. Zhang, S.-Y. Wu, An algorithm based on the generalized D-gap function for equilibrium problems. J. Comput. Appl. Math. **231**, 403–411 (2009)
137. L. Zhang, S.-Y. Wu, S.-C. Fang, Convergence and error bound of a D-gap function based Newton-type algorithm for equilibrium problems. J. Ind. Manag. Optim. **6**, 333–346 (2010)

Index

A
Anchorage condition, 79
Anchor solution, 84
Auxiliary problem, 28

C
Coercivity, 34, 35, 39, 66, 79
Complementarity problem, 3, 11
Convexity
γ-concave function, 36, 45, 62, 90
convex function, 18
τ-convex function, 20, 25, 29, 36, 43, 51, 80
convex set, 17
quasiconvex function, 18, 34, 77

D
Descent algorithm, 59, 91

E
Error bound, 43, 68
Existence of solutions, 31, 77
Extragradient algorithm, 54
hyperplane extragradient algorithm, 56, 85

F
Fan-KKM, 33
Fixed point
algorithm, 51, 81
Brouwer theorem, 108
problem, 10, 14, 24, 76

I
Inverse optimization, 11, 15

K
KKM, 32, 33
KKT conditions, 88, 106

L
Lagrangian function, 8, 88
Lipschitz behaviour of minima, 81, 94

M
Mean value theorem, 27
Merit function, 25, 59
D-gap function, 29, 45, 62
gap function, 24, 42
regularized gap function, 29, 44, 45, 61
Michael selection theorem, 112
Minty inequality, 30, 54, 67
Monotonicity
cyclically antimonotone bifunction, 37
μ-monotone bifunction, 23, 36, 44, 52, 83
μ-monotone map, 22, 63
μ-pseudomonotone bifunction, 23, 36, 42, 45, 53, 83
μ-pseudomonotone map, 22
Moving set, 81
Multivalued map, 24, 32, 39
closed, 40, 110
lower semicontinuous, 76, 77, 109
upper semicontinuous, 26, 76, 109

© Springer Nature Switzerland AG 2019   119
G. Bigi et al., *Nonlinear Programming Techniques for Equilibria*, EURO Advanced
Tutorials on Operational Research, https://doi.org/10.1007/978-3-030-00205-3

N
Nash equilibrium, 4, 12, 14
    generalized Nash equilibrium, 74
Nikaido–Isoda bifunction, 12, 15, 74

P
Pareto
    optimal, 75
    weak optimal, 7, 13
Portfolio selection, 6, 13, 75
Projection, 20, 57, 66, 86
Proximal point algorithm, 66

R
Rate of convergence, 52, 53, 55, 83

S
Saddle point, 8, 14, 88
Simplex, 32, 102
Sperner theorem, 32, 103
Stability, 39
Stopping criteria, 68

Subgradient, 19

T
Tikhonov–Browder algorithm, 65
Tikhonov well-posedness, 42
Traffic network
    with elastic demand, 74
    with fixed demand, 4

U
Unique solution, 35, 40, 44, 52, 53, 83

V
Variable feasible region, 73
Variational inequality, 6, 12, 23, 28, 53, 54, 62,
        65
    quasi-variational inequality, 75

W
Wardrop equilibrium, 5, 74

FSC
www.fsc.org

MIX

Papier | Fördert
gute Waldnutzung

FSC® C083411

Zeitfracht Medien GmbH
Ferdinand-Jühlke-Straße 7
99095 Erfurt, Deutschland
produktsicherheit@kolibri360.de